# Nonlinear
# Source Separation

Nonlinear Source Separation

Luis B. Almeida

978-3-031-01398-0    paper    Almeida
978-3-031-02526-6    ebook    Almeida

DOI    10.1007/978-3-031-02526-6

A Publication in the Springer series
*SYNTHESIS LECTURES ON SIGNAL PROCESSING* Lecture #2

First Edition

# Nonlinear
# Source Separation

**Luis B. Almeida**
Instituto das Telecomunicações, Lisboa, Portugal

*SYNTHESIS LECTURES ON SIGNAL PROCESSING #2*

## ABSTRACT

The purpose of this *lecture* book is to present the state of the art in nonlinear blind source separation, in a form appropriate for students, researchers and developers. Source separation deals with the problem of recovering sources that are observed in a mixed condition. When we have little knowledge about the sources and about the mixture process, we speak of blind source separation. Linear blind source separation is a relatively well studied subject, however *nonlinear* blind source separation is still in a less advanced stage, but has seen several significant developments in the last few years.

This publication reviews the main nonlinear separation methods, including the separation of post-nonlinear mixtures, and the MISEP, ensemble learning and kTDSEP methods for generic mixtures. These methods are studied with a significant depth. A historical overview is also presented, mentioning most of the relevant results, on nonlinear blind source separation, that have been presented over the years.

## KEYWORDS

Signal Processing, Source Separation, Nonlinear blind source separation.

To Cila, Miguel and Inês

# Contents

# Acknowledgments

José Moura, the editor of this series, invited me to write this book. Without him I would not have considered it. Harri Valpola was very influential on the form in which the ensemble learning method finally is described, having led me to learn quite a bit about the MDL approach to the method. He also kindly provided the examples of application of ensemble learning, and very carefully reviewed the manuscript, having made many useful comments. Stefan Harmeling provided the examples for the TDSEP and kTDSEP methods, and made several useful suggestions regarding the description of those methods. Mariana Almeida made several useful comments on the ensemble learning section. Andreas Ziehe commented on the description of the TDSEP method. Aapo Hyvärinen helped to clarify some aspects of score function estimation. The anonymous reviewers made many useful comments. Joel Claypool, from Morgan and Claypool Publishers, was very supportive, with his attention to the progress of the manuscript and with his very positive comments. I am grateful to them all. Any errors or inaccuracies that may remain are my responsibility, not of the people who so kindly offered their help.

# Notation

In this book we shall adopt the following notational conventions, unless otherwise noted:

- Scalar nonrandom variables are denoted by lowercase lightface letters, such as $x$.

- Nonrandom vectors are denoted by lowercase boldface letters, such as $\mathbf{x}$. A component of this vector would be $x_1$, for example.

- Scalar random variables are denoted by uppercase lightface letters, such as $X$. A specific value of the random variable will normally be denoted by the corresponding lowercase letter, $x$.

- Random vectors are denoted by boldface uppercase letters, e.g. $\mathbf{X}$. A specific value of the random vector will normally be denoted by the corresponding boldface lowercase letter, $\mathbf{x}$.

- Vectors are considered to be represented by column matrices, unless otherwise noted.

- Matrices (except for column matrices used to represent vectors) are represented by uppercase lightface letters, such as $A$. This overlaps with the notation for scalar random variables, but the context will make the meaning clear.

- We shall make a slight abuse of notation in the representation of probability density functions (pdfs), denoting them all by $p(\cdot)$, to make expressions simpler. For example, the joint density of the random variables $X_1$ and $X_2$ would normally be represented by $p_{X_1,X_2}(x_1, x_2)$, but we shall represent it simply by $p(x_1, x_2)$. We shall drop the subscripts indicating which specific random variables are meant, because the arguments that are used will make that clear, eliminating any possibility of confusion. In the same spirit, for example, the conditional density $p_{X_1 \mid X_2}(x_1 \mid x_2)$ will be denoted simply as $p(x_1 \mid x_2)$.

- The term *sample* has two different common uses. In the statistical literature it represents a set of observations of the same random variable, while in the signal processing literature it represents a single observed value of a signal, be it random or deterministic. We shall follow the signal processing convention. Therefore, for us, a sample of the signal $x(t)$ will be the value of $x(3)$, for example.

Table 1 provides a summary of the symbols and acronyms used in the book.

| **TABLE 1:**  Symbol Table | |
| --- | --- |
| *Scalars, vectors, matrices, random variables* | |
| $a, s, x$ | (lightface, lowercase): Scalars |
| $\boldsymbol{a}, \boldsymbol{s}, \boldsymbol{x}$ | (boldface, lowercase): Vectors |
| $A, S, X$ | (lightface, uppercase): Matrices; also scalar random variables |
| $\boldsymbol{A}, \boldsymbol{S}, \boldsymbol{X}$ | (boldface, uppercase): Random vectors |
| $A^{\mathrm{T}}$ | The transpose of matrix $A$ |
| $A^{-\mathrm{T}}$ | The transpose of the inverse of matrix $A$ |
| $\det A$ | The determinant of matrix $A$ |
| *Variables and functions related to mixtures and separation* | |
| $\boldsymbol{S}, \boldsymbol{s}$ | Source vectors |
| $\boldsymbol{X}, \boldsymbol{x}$ | Mixture vectors |
| $\boldsymbol{Y}, \boldsymbol{y}$ | Vectors of separated components |
| $\boldsymbol{Z}, \boldsymbol{z}$ | Auxiliary output vectors, in INFOMAX and MISEP; also the prewhitened mixture, in SOBI and TDSEP |
| $A$ | Mixture matrix |
| $W$ | Separating matrix |
| $\mathcal{M}(\cdot)$ | Nonlinear mixture function |
| $\mathcal{F}(\cdot)$ | Nonlinear separating function |
| $B$ | Prewhitening matrix, in SOBI and TDSEP |
| $Q$ | Rotation matrix, in SOBI and TDSEP |
| $\varphi(\cdot)$ | Score function, in ICA methods; also the moment generating function of a statistical distribution |
| $\psi(\cdot)$ | Output nonlinearities, in INFOMAX and MISEP; also the cumulant generating function of a statistical distribution |
| $\mathcal{X}$ | Mixture space, in kTDSEP |
| $\hat{\mathcal{X}}$ | Feature space, in kTDSEP |
| $\tilde{\mathcal{X}}$ | Intermediate space, in kTDSEP |
| *Statistical functions and operators* | |
| $p(\boldsymbol{x})$ | Probability density function (pdf) of the random variable $X$ |
| $F_X(\cdot)$ | Cumulative distribution function (CDF) of the random variable $X$ |

**TABLE 1:** Symbol Table (*Continued*)

| | |
|---|---|
| $E(X)$ | Expected value of the random variable $X$ |
| $\langle x \rangle$ | Mean of $x$ in a given set, or in the time domain |
| $H(X)$ | Shannon's differential entropy of the continuous random variable $X$; also Shannon's entropy of the discrete random variable $X$ |
| $H_n(X)$ | Renyi's differential entropy of order $n$ of the random variable $X$ |
| $J(X)$ | Negentropy of the random variable $X$ |
| $I(\boldsymbol{X})$ | Mutual information of the components of the random vector $\boldsymbol{X}$ |
| $KLD(p, q)$ | Kullback–Leibler divergence of the $q$ pdf relative to the $p$ pdf |

*Acronyms and abbreviations*

| | |
|---|---|
| CDF | Cumulative distribution function (of a random variable) |
| ICA | Independent component analysis |
| i.i.d. | Independent, identically distributed (random variables) |
| KLD | Kullback–Leibler divergence |
| MDL | Minimum description length |
| MLP | Multilayer perceptron |
| NFA | Nonlinear factor analysis |
| PCA | Principal component analysis |
| pdf | Probability density function (of a random variable) |
| RBF | Radial basis function |
| SNR | Signal-to-noise ratio |
| SOM | Self-organizing map |

# Preface

Source separation deals with the problem of recovering sources that are observed in a mixed condition. When we have little knowledge about the sources and about the mixture process, we speak of blind source separation. Linear blind source separation is a relatively well studied subject, and there are some good review books on it. Nonlinear blind source separation is still in a less advanced stage, but has seen several significant developments in the last few years.

The purpose of this book is to present the state of the art in nonlinear blind source separation, in a form appropriate for students, researchers and developers. The book reviews the main nonlinear separation methods, including the separation of post-nonlinear mixtures, and the MISEP, ensemble learning and kTDSEP methods for generic mixtures. These methods are studied with a significant depth. A historical overview is also presented, mentioning most of the relevant results on nonlinear blind source separation that have been presented over the years, and giving pointers to the literature.

The book tries to be relatively self-contained. It includes an initial chapter on linear source separation, focusing on those separation methods that are useful for the ensuing study of nonlinear separation. An extensive bibliography is included. Many of the references contain pointers to freely accessible online versions of the publications.

The prerequisites for understanding the book consist of a basic knowledge of mathematical analysis and of statistics. Some more advanced concepts of statistics are treated in an appendix. A basic knowledge of neural networks (multilayer perceptrons and backpropagation) is needed for understanding some parts of the book

The writing style is intended to afford an easy reading without sacrificing rigor. Where necessary the reader is pointed to the relevant literature, for a discussion of some more detailed aspects of the methods that are studied. Several examples of application of the studied methods are included, and an appendix provides pointers to online code and data for linear and nonlinear source separation.

CHAPTER 1

# Introduction

It is common, in many practical situations, to have access to observations that are mixtures of some "original" signals, and to be interested in recovering those signals. For example, when trying to obtain an electrocardiogram of the fetus, in a pregnant woman, the fetus' signals will be contaminated by the much stronger signals from the mother's heart. When recording speech in a noisy environment, the signals recorded through the microphone(s) will be contaminated with noise. When acquiring the image of a document in a scanner, the image sometimes gets superimposed with the image from the opposite page, especially if the paper is thin. In all these cases we obtain signals that are contaminated by other signals, and we would like to get rid of the contamination to recover the original signals.

The recovery of the original signals is normally called *source separation*. The original signals are normally called *sources*, and the contaminated signals are considered to be *mixtures* of those sources. In cases such as those presented above, if there is little knowledge about the sources and about the details of the mixture process, we normally speak of *blind source separation* (BSS). In many situations, such as most of those involving biomedical or acoustic signals, the mixture process is known to be linear, to a good approximation. This allows us to perform source separation through linear operations, and we then speak of linear source separation. On the other hand, the document imaging situation that we mentioned above is an example of a situation in which the mixture is nonlinear, and the corresponding separation process also has to be nonlinear.

Linear source separation has been the object of much study in recent years, and the corresponding theory is rather well developed. It has also been the subject of some good overview books, such as [27, 55]. Nonlinear source separation, on the other hand, has been the object of research only more recently, and until now there was, to our knowledge, no overview book specifically addressing it (a former overview paper is [59]).

This book attempts to fill this gap, by providing a comprehensive overview of the state of the art in nonlinear source separation. It is intended to be used as an introduction to the topic of nonlinear source separation for scientists and students, as well as for applications-oriented people. It has been intentionally limited in size, so as to be easy to read. For most of the methods that are studied, we present their foundations and a relatively detailed overview of their operation, but we do not discuss them in full detail. Readers who want to know more about a specific method or who want to implement it should then consult the references that are indicated.

The book tries to be relatively comprehensive and self-contained. For this purpose, it includes a chapter introducing linear source separation methods, focusing especially on those that will be more relevant to the ensuing treatment of nonlinear separation. The next and main chapter studies nonlinear separation methods, starting with the so-called post-nonlinear setting, in which there are theoretical results guaranteeing source recovery, and proceeding then to the unconstrained separation methods. In these we give special focus to the three main methods which, in our view, are best developed and potentially most useful: MISEP, ensemble learning, and kTDSEP. We also give an overview of other methods that have been proposed in the literature. A final chapter tries to show the prospects for the future in this field.

Nonlinear source separation often makes use of nonlinear trainable systems, the most common of which are multilayer perceptrons (MLPs). We shall assume that the reader has a basic knowledge of these systems, of the backpropagation training method, and of gradient-based optimization in general. Basic texts on this subject are, for example, [4, 39]. We shall also assume that the reader has a basic knowledge of statistics and of linear algebra. A few somewhat more advanced statistical concepts that are useful for understanding some parts of the book are reviewed in Appendix A.

There are several software packages, and some datasets, for source separation that are available online. Some links are given in Appendix B.

## 1.1    BASIC CONCEPTS

In its simplest setting, the linear source separation problem can be stated as follows: There is an unknown random vector $S$, whose components are called *sources*. We observe exemplars of a *mixture vector* $X$ given by

$$X = AS, \tag{1.1}$$

where $A$ is a matrix. Often the numbers of components of $S$ and $X$ are assumed to be the same, implying that matrix $A$ is square, and we speak of a *square* separation problem. In source separation one is interested in recovering the source vector $S$, and possibly also the mixture matrix $A$.

Clearly, the problem cannot be solved without some additional knowledge about $S$ and/or $A$. The assumption that is most commonly made is that the components of $S$ are statistically independent from one another, but sometimes some other assumptions are made, either explicitly or implicitly. This is clarified ahead.

Several variants of the basic source separation problem exist. First of all, the number of components of the mixture $X$ may be smaller than the number of sources, and we speak of an *overcomplete* or *underdetermined* problem, or it may be larger than the number of sources, and we speak of an *undercomplete* or *overdetermined* problem.

Another variant involves the presence of additive noise. Equation (1.1) is then replaced with

$$X = AS + N,$$

where $N$ is random noise, usually assumed to be formed by independent, identically distributed (i.i.d.) samples.

Often the sources and the mixtures are functions of time, being then denoted by $S(t)$ and $X(t)$ respectively. In this situation some more variants of the source separation problem can arise. For example, the mixture matrix $A$ may also depend on time, in which case we speak of a *nonstationary* mixture. On the other hand, the mixture vector $X(t)$, at a certain time $t$, may depend not only on the sources at the same time, but also on sources at other times. Assuming that we are still in the realm of linear mixtures, the matrix $A$ will then be a matrix of linear filters. Equation (1.1) will have to be modified accordingly, involving the convolution with these filters' impulse responses, and we speak of a *noninstantaneous* or *convolutive* mixture. These concepts extend naturally to sources that are functions of two or more variables, such as images, for example.

In the case of nonlinear source separation, which is the central topic of this book, Eq. (1.1) is replaced with

$$X = \mathcal{M}(S),$$

where $\mathcal{M}(\cdot)$ represents a nonlinear mapping. We observe $X$ and wish to recover $S$. Again, this cannot be done without some further assumptions, which will be clarified ahead.

As in the linear case, the nonlinear mixture may have noise, and may be nonstationary or noninstantaneous. However, the formal treatment of these variants is not much developed to date, and in this book we shall limit ourselves, almost always, to stationary, instantaneous, noiseless mixtures. Also, in most situations, we shall consider the mixture to be square; i.e., the sizes of $X$ and $S$ will be the same, although there will be a few cases in which we shall consider different situations.

### 1.1.1 The Scatter Plot as a Tool To Depict Joint Distributions

Scatter plots are often used to depict joint distributions of random variables. It will be useful for us to develop some experience in interpreting these plots. We give some examples in Fig. 1.1. In the left column we see two examples of joint distributions of independent sources. In the upper

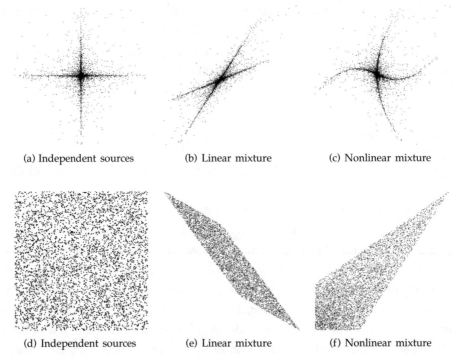

(a) Independent sources          (b) Linear mixture          (c) Nonlinear mixture

(d) Independent sources          (e) Linear mixture          (f) Nonlinear mixture

**FIGURE 1.1:** Examples of linear and nonlinear mixtures. The top row shows mixtures of two super-gaussian sources and the bottom row shows mixtures of two uniformly distributed sources

example the sources have densities that are strongly peaked (they are much more peaked than Gaussian distributions with the same variance, and for this reason they are called *supergaussian*).[1] The lower example shows uniformly distributed sources.

Note that, in both cases, a horizontal section through the distribution of the sources—Figs. 1.1(a) and 1.1(d)—will yield a density which has the same shape (apart from an amplitude scale factor) irrespective of where the section is made. The same happens with vertical sections. These sections, once normalized, correspond to conditional distributions, and the fact that their shape is independent of where the section is made means that the conditional distribution of one of the sources given the other is independent of that other source. For example, $p(s_1 \mid s_2)$ is independent of $s_2$. This means that the sources are independent from each other.

The middle column of the figure shows examples of linear mixtures. In such mixtures, lines that were originally straight remain straight, and lines that were originally parallel to each

---

[1]More specifically, a random variable is called *supergaussian* if its kurtosis is positive, and *subgaussian* if its kurtosis is negative (the kurtosis of a Gaussian random variable is zero). The kurtosis of a random variable is defined as its fourth-order cumulant (see Section 2.4.1).

other remain parallel to each other. Linear operations can only perform scaling, rotation, and shear transformations, which do not affect the straightness or the parallelism of lines. But note that in these mixtures the two components are no longer independent from each other. This can be seen by the fact that sections (either horizontal or vertical) through the distribution yield densities that depend on where the section was made.

The rightmost column shows two examples of nonlinear mixtures. In the upper one, lines that were originally straight now appear curved. In the lower one, lines that were originally parallel to each other (opposite edges of the square) do not remain parallel to each other, even though they remain straight. Both of these are telltale signs of nonlinear mixtures. Whenever they occur we know that the mixture that is being considered is nonlinear. And once again, we can see that the two random variables in each mixture are not independent from each other: The densities corresponding to horizontal or vertical sections through the distribution depend on where these sections were made.

## 1.1.2 Separation Criteria

In blind source separation, both linear and nonlinear, we normally assume that we know very little about the mixture process. We may know, for example, that it is instantaneous and invariant in time, but we usually assume that we do not know the mixture matrix $A$, in the linear case, or the mixture function $\mathcal{M}$, in the nonlinear case. The sources $S_i$ are normally considered to be unknown too. But of course some knowledge has to be assumed in order to make the problem solvable. An assumption that is very often made is that the sources are mutually statistically independent. This may seem, at first, to be a rather bold assumption. However, it often turns out to be valid in practice. For example, when dealing with a mixture of sound signals from two or more physically distinct sources (such as speech from two or more speakers, or speech and background noise) the different sound sources can be considered independent, to a good approximation. In biomedical applications, signals coming from physically separate sources, such as the cardiac signals from mother and fetus (in the electrocardiography of pregnant women), or the signals generated in different parts of the brain (in electroencephalographic applications) can often be considered independent from one another, to a good approximation. The same happens in many other situations, in fields as diverse as telecommunications and astronomy, for example.

Consequently, one of the main criteria that are used for the separation of sources is statistical independence. This creates the need to measure the statistical dependence of random variables. A measure of the statistical dependence of several random variables $Y_i$ should have two basic properties: It should be equal to zero if the random variables are mutually statistically independent, and should be positive otherwise. A measure with these properties is called a *contrast function* by Comon [28].

One of the main measures of statistical dependence used in source separation is the mutual information, denoted $I(\cdot)$. Given a random vector $\boldsymbol{Y}$, $I(\boldsymbol{Y})$ measures the amount of information that is shared by the components of $\boldsymbol{Y}$. It is zero if these components are independent and is positive if they have any mutual dependence. The minimization of $I(\boldsymbol{Y})$ therefore leads the components of $\boldsymbol{Y}$ to be as independent from one another as possible. Mutual information is defined, and some of its properties are studied, in Appendix A.4. Readers that are not familiar with this concept should read that appendix, as well as the related Appendices A.2 and A.3, in order to understand several of the linear and nonlinear separation methods discussed in this book.

## 1.2   SUMMARY

Source separation deals with the recovery of *sources* that are observed in a mixed condition. *Blind* source separation (BSS) refers to source separation in situations in which there is little knowledge about the sources and about the mixing process.

Most of the work done to date on source separation concerns linear mixtures. Variants that have been studied include over- and undercomplete mixtures, noisy, nonstationary, and convolutive (noninstantaneous) mixtures.

This book is mainly concerned with nonlinear mixtures. These may also be noisy, nonstationary, and/or noninstantaneous, but we shall normally restrict ourselves to the basic case of noise-free, stationary instantaneous mixtures, because other variants have still been an object of very little study in the nonlinear case. The book analyzes with some detail the main nonlinear separation methods, and gives an overview of other methods that have been proposed.

An assumption that is frequently made in blind source separation is that the sources are independent random variables. The mutual dependence of random variables is often measured by means of their mutual information.

CHAPTER 2

# Linear Source Separation

In this chapter we shall make a brief overview of linear source separation in order to introduce several concepts and methods that will be useful later for the treatment of nonlinear separation. We shall only deal with the basic form of the linear separation problem (same number of observed mixture components as of sources, no noise, stationary instantaneous mixture) since this is, almost exclusively, the form with which we shall be concerned later, when dealing with nonlinear separation.

## 2.1 STATEMENT OF THE PROBLEM

Consider a random vector $S$, whose components we call sources, and a mixture vector given by

$$X = AS,$$

where the sizes of $S$ and $X$ are equal, and $A$ is a square, invertible matrix, which is usually called the *mixture matrix* or *mixing matrix*. We observe $N$ exemplars of $X$, generated by identically distributed (but possibly nonindependent) exemplars of $S$. These exemplars of $S$, however, are not observed. In the *blind* separation setting, we assume that we do not know the mixing matrix, and that we have relatively little knowledge about $S$. We wish to recover the sources, i.e. the components of $S$.

If we knew the mixing matrix, we could simply invert it and compute the sources by means of the inverse

$$S = A^{-1}X.$$

However, in the blind source separation setting we do not know $A$, and therefore we have to use some other method to estimate the sources. As we have said above, the assumption that is most commonly made is that the sources (the components of $S$) are mutually statistically independent.

In accordance with this assumption, one of the most widely used methods to recover the sources consists of estimating a square matrix $W$ such that the components of $Y$, given by

$$Y = WX, \tag{2.1}$$

are mutually statistically independent. It has been shown by Comon [28] that, if at most one of the components of $S$ has a Gaussian distribution, this independence criterion suffices to recover $S$, in the following sense: If $Y$ is obtained through (2.1) and its components are mutually independent, then $Y$ and $S$ are related by

$$Y = PDS,$$

where $P$ is a permutation matrix[2] and $D$ is a diagonal matrix. This means that the components of $Y$ are the components of $S$, possibly subject to a permutation and to arbitrary scalings.

This use of the independence criterion leads to the so-called *independent component analysis* (ICA) techniques, which analyze mixtures into sets of components that are as independent from one another as possible, according to some mutual dependence measure.

Note that simply imposing that the components of $Y$ be uncorrelated with one another does not suffice for separation. There is an infinite number of solutions of (2.1) in which the components of $Y$ are mutually uncorrelated, but in which each component still contains contributions from more than one source.[3] Also note that, for the same random vector $X$, principal component analysis (PCA) [39] and independent component analysis normally yield very different results. Although both methods yield components that are mutually uncorrelated, the components extracted by PCA normally are not statistically independent from one another, and normally contain contributions from more than one source, contrary to what happens with ICA.

There are several practical methods for estimating the matrix $W$ based on the independence criterion (see [27, 55] for overviews). In the next sections we shall briefly describe some of them, focusing on those that will be useful later in the study of nonlinear separation.

## 2.2    INFOMAX

INFOMAX, also often called the *Bell–Sejnowski method*, is a method for performing linear ICA; i.e., it attempts to transform the mixture $X$, according to (2.1), into components $Y_i$ which are as independent from one another as possible.

The INFOMAX method is based on the structure depicted in Fig. 2.1. In the left-hand side of the figure we recognize the implementation of the separation operation (2.1). The result of the separation is $Y$. The $\psi_i$ blocks and the $Z_i$ outputs are auxiliary, being used only during the optimization process.

---

[2]A permutation matrix has exactly one element per row and one element per column equal to 1, all other elements being equal to zero.

[3]However, decorrelation, in a different setting, can be a criterion for separation, as we shall see in Section 2.3.

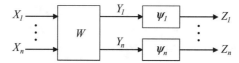

**FIGURE 2.1:** Structure used by INFOMAX. The $W$ block performs a product by a matrix, and is what performs the separation proper. The separated outputs are $Y_i$. The $\psi_i$ blocks, implementing nonlinear increasing functions, are auxiliary, being used only during optimization

In the paper that introduced it [19], INFOMAX was justified on the basis of an information maximization criterion (hence its name). However, the method has later been interpreted as a maximum likelihood method [83] and also as a method based on the minimization of the mutual information $I(Y)$ [55] (recall that mutual information was seen, in Section 1.1.2, to be a measure of statistical dependence). Here we shall use the interpretation based on the minimization of mutual information, because this will be the most useful approach when we deal with nonlinear separation methods, later on.

For presenting the mutual information interpretation, we shall start by assuming that each of the $\psi_i$ functions is equal to the cumulative distribution function (CDF) of the corresponding random variable $Y_i$ (we shall denote that cumulative function by $F_{Y_i}$). Then, all of the $Z_i$ will be uniformly distributed in (0, 1) (see Appendix A.1). Therefore, $p(z_i)$ will be 1 in (0, 1) and zero elsewhere, and the entropy of each of the $Z_i$ shall be zero, $H(Z_i) = 0$.

As shown in Appendix A.4, the mutual information is not affected by performing invertible, possibly nonlinear, transformations on the individual random variables. In the above setting, since the $\psi_i$ functions are invertible, this means that $I(Y) = I(Z)$. Therefore,

$$
\begin{aligned}
I(Y) &= I(Z) \\
&= \sum_i H(Z_i) - H(Z) \\
&= -H(Z).
\end{aligned}
\tag{2.2}
$$

This result shows that minimizing the mutual information of the components of $Y$ can be achieved by maximizing the output entropy, $H(Z)$.[4] This is an advantage, since the direct minimization of the mutual information is generally difficult, and maximizing the output entropy is easier, as we shall see next.

The output entropy of the system of Fig. 2.1 is related to the input entropy by[5]

$$
H(Z) = H(X) + \mathrm{E}[\log|\det J|],
$$

---

[4]We are considering continuous random variables, and therefore $H(\cdot)$ represents the differential entropy, throughout this and the next chapter (see Appendices A.2 and A.4). We shall designate it simply as *entropy*, for brevity.

[5]See Appendix A.2.1.

where $E(\cdot)$ denotes statistical expectation and $J = \partial Z/\partial X$ is the Jacobian of the transformation performed by the system. $H(X)$ does not depend on $W$, and therefore the maximization of $H(Z)$ is equivalent to the maximization of $E[\log |\det J|]$.

To perform that maximization we need to have a set of observations of $X$ (often called the *training set*) from which we can compute the corresponding set of outputs $Z$. We approximate the statistical expectation by the mean in this set:

$$E[\log |\det J|] \approx \frac{1}{M} \sum_{m=1}^{M} \log |\det J_m| = \mathcal{J},$$

where $J_m$ is the Jacobian corresponding to the $m$th exemplar of $X$ in the training set (the $m$th training pattern) and $M$ is the number of exemplars in the training set. $\mathcal{J}$ is what shall effectively be used as objective function to be maximized.

The maximization is usually done by means of gradient-based methods. We have

$$\mathcal{J} = \frac{1}{M} \sum_{m=1}^{N} \mathcal{L}_m \qquad (2.3)$$

with

$$\mathcal{L}_m = \log |\det J_m| . \qquad (2.4)$$

For simplicity we shall drop, from now on, the subscript $m$, which refers to the training pattern being considered. The Jacobian in the preceding equation is given by

$$J = W \prod_j \psi'_j(y_j).$$

The $\psi_j$ functions are normally chosen as increasing functions. Therefore $\psi'_j(y_j) \geq 0$ and

$$\mathcal{L} = \log |\det W| + \sum_j \log \psi'_j(y_j).$$

The gradient relative to $W$ is given by

$$\frac{\partial \mathcal{L}}{\partial W} = W^{-\mathrm{T}} + \sum_j \frac{\partial}{\partial W} \log \psi'_j(y_j)$$

$$= W^{-\mathrm{T}} + \sum_j \frac{\psi''_j(y_j)}{\psi'_j(y_j)} \frac{\partial y_j}{\partial W},$$

where $W^{-\mathrm{T}}$ is the transpose of the inverse of $W$.

In the preceding equation, $\partial y_j/\partial W$ is a matrix whose $j$th row is $x^{\mathrm{T}}$ (the superscript T denoting matrix transposition), and all other rows are equal to zero. Therefore, if we define the vector

$$\boldsymbol{\xi} = \left[ \frac{\psi_1''(y_1)}{\psi_1'(y_1)}, \frac{\psi_2''(y_2)}{\psi_2'(y_2)}, \cdots, \frac{\psi_n''(y_n)}{\psi_n'(y_n)} \right]^{\mathrm{T}}, \qquad (2.5)$$

where $n$ is the size of $\boldsymbol{y}$, we finally have

$$\frac{\partial \mathcal{L}}{\partial W} = W^{-\mathrm{T}} + \boldsymbol{\xi} \boldsymbol{x}^{\mathrm{T}}. \qquad (2.6)$$

This expression of the gradient is the one to be used in the *stochastic* or *online* optimization mode [4, 39], in which the gradient of $\mathcal{L}$ is used to update $W$ after the presentation of each training pattern at the input of the system of Fig. 2.1. An alternative is to use the so-called *deterministic* or *batch* optimization mode, in which the gradient of $\mathcal{J}$ is used to update $W$. For that mode we have

$$\frac{\partial \mathcal{J}}{\partial W} = W^{-\mathrm{T}} + \langle \boldsymbol{\xi} \boldsymbol{x}^{\mathrm{T}} \rangle, \qquad (2.7)$$

where $\langle \cdot \rangle$ denotes the mean computed in the training set. Since this gradient involves a mean in the training set, the update of $W$ is performed once after each sweep through the whole training set, in this mode.

These equations allow us to compute the gradient of the objective function, for use in any gradient-based minimization method. The methods that are more commonly used for minimization are simple gradient descent and, more frequently, the so-called *relative gradient* or *natural gradient* method (see [13, 25, 55]). The relative/natural gradient method has the double advantage of being computationally simpler, avoiding the matrix inversion that appears in (2.6) and (2.7), and being usually faster, needing fewer iterations for convergence. As shown in the indicated references, this method yields the update equation

$$\Delta W \propto (I + \boldsymbol{\xi} \boldsymbol{y}^{\mathrm{T}}) W,$$

in the stochastic mode, and

$$\Delta W \propto (I + \langle \boldsymbol{\xi} \boldsymbol{y}^{\mathrm{T}} \rangle) W$$

in the deterministic mode.

## 2.2.1   Choice of the Output Nonlinearities

One aspect that we have not discussed yet is how to choose the $\psi_i$ nonlinearities. As previously mentioned, they should ideally be the CDFs of the corresponding random variables, $\psi_i = F_{Y_i}$. In practice, however, these cumulative functions often are not known, or are only known in a qualitative or a very coarse form. Fortunately it is normally not necessary to use a very good approximation of these cumulative functions to obtain good results with linear separation. The reason is that linear ICA is a rather constrained problem. For example, when there are two sources to be separated, there are four scalar unknowns in matrix $W$. However, after optimization, there will still remain two arbitrary scale factors, one for each extracted component. Therefore the problem effectively corresponds to finding just two scalars, and is consequently very constrained.

The fact that the problem is strongly constrained makes it possible to obtain good results even with rather crude approximations of the CDFs. For example, it is known that if the source distributions are approximately symmetric and more peaked than a Gaussian (i.e. supergaussian) distribution, sigmoidal nonlinearities of the form

$$\psi(y) = \frac{1}{1 + e^{-y}}$$

will suffice for a nearly perfect separation [19]. It has been proposed by some authors to use just one of two predetermined $\psi$ functions, one for supergaussian and one for subgaussian components, as a means to encompass the large majority of practical situations [66].

Often the choice of the $\psi_i$ nonlinearities turns out not to be too difficult when applying INFOMAX to linear separation problems, especially if we have some qualitative knowledge about the sources (for example, if we know that they are speech signals, or that they are images). In nonlinear separation, however, the choice of these nonlinearities will become much more stringent, because nonlinear separation is far less constrained than linear separation, and a poor estimation of the CDFs can easily lead to poor separation. It is therefore useful to examine methods for a more accurate estimation of these functions.

It is always possible to estimate the densities $p(y_i)$ by one of several standard procedures (for example using kernel estimators, or using mixtures of Gaussians estimated with the EM algorithm [84]). It is then possible to set $\psi_i$ equal to the corresponding CDFs. Although this has been done in some cases [22, 109], it seems preferable to use a form of estimation which has a closer relationship with the optimization that is being performed in the ICA process. In the next subsections we shall describe two methods that have been proposed for estimating these nonlinearities in the context of linear ICA. Both of these methods will later be relevant for the study of nonlinear source separation.

### 2.2.2    Maximum Entropy Estimation of the Nonlinearities

This method for estimating the $\psi_i$ nonlinearities (proposed by L. Almeida [6]) uses, as a criterion for estimating these nonlinearities, the maximization of the output entropy $H(\mathbf{Z})$. This is the same criterion that is used for optimization of the separating matrix $W$ itself. Our presentation of the method will be somewhat detailed because this estimation technique, together with the INFOMAX method for estimating the separating matrix, constitutes a good introduction to the MISEP nonlinear separation method, to be presented later in Section 3.2.1 (and, in fact, constitutes the linear version of MISEP). Once this linear version is understood, the extension to nonlinear separation will be relatively easy to grasp.

### 2.2.2.1 Theoretical Basis

Assume that we represent the $\psi_i$ nonlinearities by some parameterized estimates $\hat{\psi}_i(y_i, \boldsymbol{w}_i)$, where $\boldsymbol{w}_i$ are vectors of parameters. These functions are chosen to be nondecreasing functions of $y_i$, with codomain $(0, 1)$. Also assume, for the moment, that $W$ is kept fixed, so that the distribution of $\boldsymbol{Y}$ is also fixed. From (2.2),

$$H(\boldsymbol{Z}) = \sum_i H(Z_i) - I(\boldsymbol{Y}). \qquad (2.8)$$

The distribution of $\boldsymbol{Y}$ has been assumed to be fixed, and therefore $I(\boldsymbol{Y})$ is constant. Maximization of $H(\boldsymbol{Z})$ thus corresponds to the maximization of $\sum_i H(Z_i)$. Since the parameter vectors $\boldsymbol{w}_i$ can be varied independently from one another, this corresponds to the simultaneous maximization of all the $H(Z_i)$ terms.

Given that $\hat{\psi}_i$ has codomain $(0, 1)$, $Z_i$ is limited to that interval. We therefore have, for each $Z_i$, the maximization of the entropy of a continuous random variable, constrained to the interval $(0, 1)$. The maximum entropy distribution under this constraint is the uniform distribution (see Appendix A.2). Therefore the maximization will lead each $Z_i$ to become as close as possible to a uniformly distributed variable in $(0, 1)$, subject only to the restrictions imposed by the limitations of the family of functions $\hat{\psi}_i$.[6]

Since, at the maximum, $Z_i$ is approximately uniformly distributed in $(0, 1)$, and since $\hat{\psi}_i(y_i, \boldsymbol{w}_i)$ is, by construction, a nondecreasing function of $y_i$, it must be approximately equal to the CDF $F_{Y_i}$, as desired (see Appendix A.1). Therefore, maximization of $H(\boldsymbol{Z})$ leads the $\hat{\psi}_i(y_i, \boldsymbol{w}_i)$ functions to approximate the corresponding CDFs, subject only to the limitations of the family of approximators $\hat{\psi}_i(y_i, \boldsymbol{w}_i)$.

Let us now drop the assumption that $W$ is constant, and assume instead that it is also being optimized by maximization of $H(\boldsymbol{Z})$. Consider what happens at a maximum of $H(\boldsymbol{Z})$. Each of the $H(Z_i)$ must be maximal, because otherwise $H(\boldsymbol{Z})$ could still be increased further, by increasing that specific $H(Z_i)$ while keeping the other $Z_j$ and $Y$ fixed—see Eq. (2.8). Therefore, at a maximum of $H(\boldsymbol{Z})$, the $Z_i$ variables must be as uniform as possible, and the $\hat{\psi}_i$ functions must be the closest possible approximations of the corresponding CDFs.

This shows that the maximization of the output entropy will lead to the desired estimation of the CDFs by the $\hat{\psi}_i$ functions while at the same time optimizing $W$. In practice, during

---

[6]We are disregarding the possibility of stopping at a local maximum during the optimization. That would be a contingency of the specific optimization procedure being used, and not of the basic method itself.

the optimization process, when $W$ changes rapidly, the $\hat{\psi}_i$ functions will follow the corresponding CDFs with some lag, because the CDFs are changing rapidly. During the asymptotic convergence to the maximum of $H(\mathbf{Z})$, $W$ will change progressively slower, and the $\hat{\psi}_i$ functions will approximate the corresponding CDFs progressively more closely, tending to the best possible approximations.

In summary, the maximization of the output entropy leads to the minimization of the mutual information $I(\mathbf{Y})$, with the simultaneous adaptive estimation of the $\psi_i$ nonlinearities.

Before proceeding to discuss the practical implementation of this method for the estimation of the nonlinearities, we note that it is based on the optimization of the same objective function that is used for the estimation of the separating matrix. Therefore, the whole optimization is based on a single objective function. There are maximization methods (e.g. methods based on gradient ascent [4]) which are guaranteed to converge when there is a single function to be optimized, which is differentiable and upper-bounded, as in this case. Therefore there is no risk of instability, contrary to what happens if the $\psi_i$ nonlinearities are estimated based on some other criterion.

### 2.2.2.2 Implementation of the Method

We have given an explanation of the theoretical principles behind the estimation of the $\psi_i$ functions, and we shall now describe how these principles are put to practice. Each of the estimates $\hat{\psi}_i(y_i, \boldsymbol{w}_i)$ is implemented by means of a multilayer perceptron (MLP) with a single input and a single output. The $\boldsymbol{w}_i$ parameter vector is formed by the MLP's weights. Each MLP has a linear output unit and a single hidden layer with a number of units that depends on the specific problem being dealt with, but which is typically between 2 and 20.

The whole structure of Fig. 2.1, with the $\psi_i$ blocks implemented by means of MLPs, can then be seen as a single, larger multilayer perceptron with two hidden layers. The first hidden layer's units are linear, corresponding to the output units of block $W$, which have as outputs $Y_i$. The second hidden layer is nonlinear, and is formed by the hidden layers of all the $\psi_i$ blocks taken together. The output layer is linear, being formed by the output units of all the $\psi_i$ blocks taken together.

Figure 2.2 depicts this MLP in a form that will be useful for the discussion that follows. The input is the mixture vector $\boldsymbol{x}$. The $W$ block performs a product (on the left) by matrix $W$.[7] Its outputs form the vector of extracted components, $\boldsymbol{y}$. In the figure we consider this vector augmented with a component equal to 1, which is useful for implementing the bias terms of the following layer's units, and we denote the augmented vector by $\bar{\boldsymbol{y}}$.

---

[7]We shall designate each linear block by the same letter that denotes the corresponding matrix, because this will not cause any confusion.

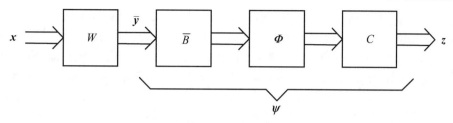

**FIGURE 2.2:** Structure of the INFOMAX system with adaptive estimation of the nonlinearities. The part marked $\psi$ corresponds to all the $\psi_i$ blocks taken together. See the text for further explanation

The three rightmost blocks represent all the $\psi_i$ blocks of Fig. 2.1 taken together. Block $\bar{B}$ performs a product by matrix $\bar{B}$, which is the weight matrix of the hidden units of all the $\psi_i$ blocks, taken as forming a single hidden layer. As in $\bar{y}$, the overbar in $\bar{B}$ indicates that this matrix incorporates the elements relative to the bias terms of the hidden units. The output of this block consists of the vector of input activations of the units of the global hidden layer of all the $\psi_i$ blocks.

Block $\Phi$ is nonlinear and applies to its input vector, on a per-component basis, the nonlinear activation functions of the units of that hidden layer. We shall denote these activation functions by $\phi_i$. Usually these activation functions are the same for all units, normally having a sigmoidal shape. A common choice is the hyperbolic tangent function.

The output of the $\Phi$ block is the vector of output activations of the global hidden layer. Block $C$ multiplies this vector, on the left, by matrix $C$, which is the weight matrix of the linear output units of the $\psi_i$ blocks. The output of the $C$ block is the output vector $z$.

Of course, since there are no interconnections between the various $\psi_i$ blocks, matrices $\bar{B}$ and $C$ have a special structure, with a large number of elements equal to zero. This does not affect our further analysis, except for the fact that those elements are never changed during the optimization, being always kept equal to zero, to keep the correct network structure.

We wish to compute the gradient of the output entropy $H(Z)$ relative to the weights of this network, so that we can use gradient-based maximization methods. We shall not give explicit expressions of the components of the gradient here, because they are somewhat complex (and would become even more complex later, in the nonlinear MISEP method). Instead, we shall derive a method for computing these gradient components, based on the backpropagation method for multilayer perceptrons [4, 39]. As normally happens with backpropagation, this method is both more compact and more efficient than a direct computation of each of the individual components of the gradient.

Let us take Eq. (2.4), that we reproduce here:

$$\mathcal{L}_m = \log \left| \det J_m \right|.$$

We recall that $\mathcal{L}_m$ is the term of the objective function that refers to the $m$th training pattern. The objective function is the mean of all the $\mathcal{L}_m$ terms, and therefore its gradient is the mean of the gradients of all these terms. These gradients are what we now wish to compute. For simplicity we shall drop, once again, the subscript $m$, which refers to the specific training pattern being considered:

$$\mathcal{L} = \log \left| \det J \right|.$$

This function depends on the Jacobian

$$J = \frac{\partial z}{\partial x}.$$

### 2.2.2.3  A Network for Computing the Jacobian

To be able to use backpropagation for computing the gradient of $\mathcal{L}$, we shall construct a network that computes $J$, and shall then backpropagate through it, as explained ahead. Since $J$ is a matrix of first derivatives, this network is essentially a linearization of the network of Fig. 2.2. To understand how this network is built, we start by noting that the desired Jacobian is given by

$$J = C \Phi' B W, \tag{2.9}$$

where $\Phi'$ is a diagonal matrix whose diagonal elements are $\phi_i'$, i.e. the derivatives of the activation functions of the corresponding hidden units, for the specific input activations that they receive for the input pattern being considered. $B$ is the weight matrix $\bar{B}$ stripped of the column corresponding to the bias weights (those weights disappear from the equations when differentiating relative to $x$).

The network that computes $J$ according to this equation is shown in Fig. 2.3. The lower part is what computes (2.9) proper. It propagates matrices (this is depicted in the figure by the "3-D arrows"). Its input is the identity matrix $I$ of size $n \times n$ ($n$ being the number of sources and of mixture components). Block $W$ performs a product, on the left, by matrix $W$, yielding matrix $W$ itself at the output (this might seem unnecessary but is useful later, when backpropagating, to allow the computation of derivatives relative to the elements of $W$). The following blocks also perform products, on the left, by the corresponding matrices.

It is clear that the lower chain computes $J$ as per (2.9). The upper part of the network is needed because the derivatives in the diagonal matrix $\Phi'$ depend on the input pattern. The two leftmost blocks of the upper part compute the input activations of the nonlinear hidden units, and transmit these activations, through the gray-shaded arrow, to block $\Phi'$, allowing it to compute the activation function derivatives. Supplying this information is the reason for the presence of the upper part of the network.

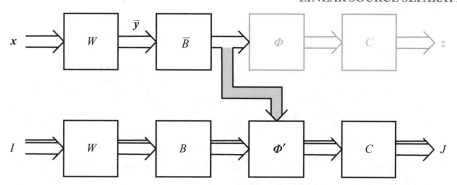

**FIGURE 2.3:** Network for computing the Jacobian. The lower part is what computes the Jacobian proper, and is essentially a linearized version of the network of Fig. 2.2. The upper part is identical to the network of Fig. 2.2. The two rightmost blocks, shown in light gray, are not necessary for computing $J$, and are shown only for better correspondence with Fig. 2.2. See the text for further explanation

### 2.2.2.4 Backpropagating to Compute the Gradient

It is known, from the theory of neural networks [4], that if we have a network with output vector $\boldsymbol{u}$ and wish to compute the gradient of a function $C(\boldsymbol{u})$ relative to the network's weights, we can do so by standard backpropagation, using as input to the backpropagation network the vector $\partial C/\partial \boldsymbol{u}$. We shall not prove that fact here, nor shall we give details of the procedure. The description and the justification can be found in [4], and in several other standard texts on neural networks.

In our case we have a network that computes $J$, and wish to compute the gradient of the function

$$\mathcal{L} = \log \left| \det J \right|.\qquad(2.10)$$

We can, therefore, use the standard backpropagation procedure.[8] The inputs to the backpropagation network are

$$\frac{\partial \mathcal{L}}{\partial J} = J^{-\mathrm{T}},$$

where the $^{-\mathrm{T}}$ superscript denotes the transpose of the inverse of the matrix.

The backpropagation method to be used is rather standard, except for two aspects that we detail here:

- All the blocks that appear in the network of Fig. 2.3 are standard neural network blocks, with the exception of block $\Phi'$. We shall now see how to backpropagate through

---

[8]The fact that the network's outputs form a matrix instead of a vector is unimportant: we can form a vector by arranging all the matrix elements into a single column.

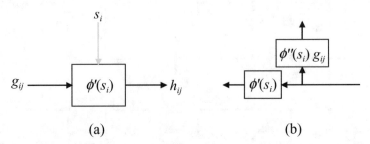

(a)                                      (b)

**FIGURE 2.4:** Backpropagation through the $\Phi'$ block. (a) Forward unit. (b) Backpropagation unit. Each box denotes a product by the value indicated inside the box

this block. A unit of this block, i.e. a unit that performs the product by the derivative of a hidden unit's activation function, is shown in Fig. 2.4(a). The unit receives an input (that we denote by $g_{ij}$) from block $B$, and receives the activation value $s_i$ (which is necessary to compute the derivative) from the upper part of the network, through the gray-shaded arrow of Fig. 2.3. This unit produces the output (that we denote by $h_{ij}$)

$$h_{ij} = \phi'(s_i)g_{ij}.$$

That output is sent to block $C$.

The backpropagation has to be made through all signal paths, and therefore has to be made to the left toward block $B$, but also upward, through the gray-shaded arrow, toward block $\bar{B}$. It is driven by the following equations:

$$\frac{\partial h_{ij}}{\partial g_{ij}} = \phi'(s_i)$$

$$\frac{\partial h_{ij}}{\partial s_i} = \phi''(s_i)g_{ij}.$$

Figure 2.4(b) shows a unit that implements this backpropagation.

- The network of Fig. 2.3 has what is called, in neural network parlance, *shared weights*; i.e., it has several connections with weights that must always remain equal to one another. In fact, the lower part of the network, propagating matrices, can be seen as $n$ identical networks, each propagating one of the column vectors of matrix $I$. All these networks share matrices $W$, $B$, and $C$. Furthermore, these networks share $W$ and $B$ with the upper part of the network. When applying backpropagation, the appropriate procedure for computing the partial derivatives relative to shared weights has to be used: The partial derivative relative to each instance of a shared weight is computed. Then, the derivatives relative to all the instances of the same weight are added together to form the partial derivative relative to that shared weight [4].

### 2.2.2.5  Constraining the MLPs

As was said in the theoretical presentation of the method, it is assumed that the $\psi_i$ blocks of Fig. 2.1 are constrained to implement nondecreasing functions with values in $(0, 1)$. In practical realizations the latter interval is normally changed to $(-1, 1)$. This does not change anything fundamental in the reasoning. The only important change is that these blocks will now estimate a scaled CDF, $2F_{Y_i} - 1$, instead of $F_{Y_i}$ itself. But as one can easily check, the whole system will still minimize the mutual information $I(Y)$. On the other hand, this change has the advantage that it allows the use of bipolar sigmoids (e.g. *tanh* functions) in the MLPs, and this kind of sigmoid is known to lead to faster optimization [64].

To implement the constraint of the MLPs' outputs to $(-1, 1)$, several methods are possible, but the one that has shown to work best in practice consists of the following:

- The activation functions of the hidden layer's units are chosen to be sigmoids with values in $(-1, 1)$, such as *tanh* functions.
- The vector of weights leading to each linear output unit is normalized, after each gradient update, to an Euclidean norm of $1/\sqrt{h}$, where $h$ is the number of hidden units feeding that output unit.

This guarantees that the MLPs' outputs will always be in the interval $(-1, 1)$. The constraint to nondecreasing functions can also be implemented in several ways. The one that has shown to be best is in fact a soft constraint:

- The hidden units' sigmoids are all chosen to be increasing functions (e.g. *tanh*).
- All the MLP's weights are initialized to positive values (with the exception of the hidden units' biases, which may be negative).

In strict terms, this only guarantees that, upon initialization, the MLPs will implement increasing functions. However, during the optimization process, the weights will tend to stay positive because the optimization maximizes the output entropy, and any sign change in a weight would tend to decrease the output entropy. In practice it has been found that negative weights occur very rarely in the optimizations, and when they occur they normally switch back to positive after a few iterations.

**An Example**    We shall now present an example of the use of INFOMAX, with entropy-based estimation of the nonlinearities, for the separation of two sources. One of the sources was strongly supergaussian, while the other was uniformly distributed (and therefore subgaussian). Figure 2.5(a) shows the joint distribution of the sources.

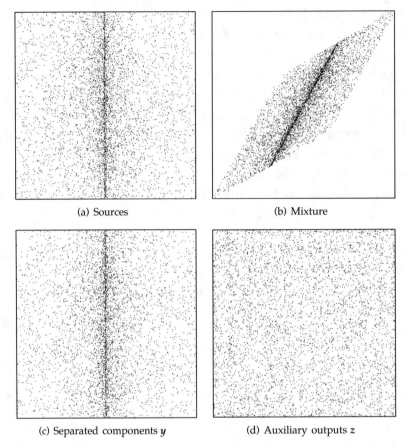

(a) Sources

(b) Mixture

(c) Separated components $y$

(d) Auxiliary outputs $z$

**FIGURE 2.5:** Scatter plots corresponding to the example of linear separation by INFOMAX with maximum entropy estimation of the output nonlinearities

The mixture matrix was

$$A = \begin{bmatrix} 1.0 & 0.4 \\ 0.8 & 1.0 \end{bmatrix}.$$

Figure 2.5(b) shows the mixture distribution. Figure 2.5(c) shows the result of separation, and we can see that it was virtually perfect. This can be confirmed by checking the product of the separating matrix that was estimated, $W$, with the mixing matrix

$$WA = \begin{bmatrix} 8.688 & 0.002 \\ -0.133 & 8.114 \end{bmatrix}.$$

This product is very close to a diagonal matrix, confirming the quality of the separation.

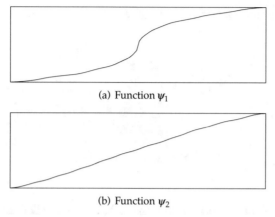

(a) Function $\psi_1$

(b) Function $\psi_2$

FIGURE 2.6: Cumulative functions learned by maximum entropy estimation

The MLPs that were used to estimate the cumulative distributions had 10 hidden units each. Figure 2.6 shows the functions that were estimated by these networks. We see that they match well (qualitatively, at least) the cumulative distributions of the corresponding sources. The quality of the estimation can be checked better by observing the distribution of the auxiliary output $z$, shown in Fig. 2.5(d). We can see that this distribution is nearly perfectly uniform within a square. On the one hand, this means that the cumulative functions were quite well estimated. On the other hand, it also confirms that the extracted components were virtually independent from each other, because otherwise $z$ could not have a distribution with this form.

This example illustrates both the effectiveness of INFOMAX in performing linear separation and the effectiveness of the maximum entropy estimation of the cumulative distributions. As we mentioned above, INFOMAX with this estimation method corresponds to the linear version of MISEP, whose extension to nonlinear separation will be studied in Section 3.2.1.

## 2.2.3    Estimation of the Score Functions

From Eqs. (2.5) and (2.6) we see that, in INFOMAX, the output nonlinearities appear only through the quotients of derivatives

$$\hat{\varphi}_j(y_j) = \frac{\psi_j''(y_j)}{\psi_j'(y_j)}.$$

We can therefore work directly with the functions $\hat{\varphi}_j$ instead of the functions $\psi_j$. Furthermore we know that, ideally, $\psi_j(y_j) = F_{Y_j}(y_j)$, where $F_{Y_j}$ is the CDF of $Y_j$. Therefore the objective of the estimation of the nonlinearities is to make $\hat{\varphi}_j(y_j)$ as close as possible to

$$
\begin{aligned}
\varphi_j(y_j) &= \frac{F''_{Y_j}(y_j)}{F'_{Y_j}(y_j)} \\
&= \frac{p'(y_j)}{p(y_j)} \\
&= \frac{\mathrm{d}}{\mathrm{d}y_j} \log p(y_j).
\end{aligned}
\tag{2.11}
$$

These $\varphi_j$ functions play an important role in ICA, and are called *score functions*. Since their definition involves pdfs, one would expect that their estimation would also involve an estimation of the probability densities. However, in [97] Taleb and Jutten proposed an interesting estimation method that involves the pdfs only indirectly, through expected values that can easily be estimated by averaging on the training set. This is the method that we shall now study.

For the derivation of this method we shall drop, for simplicity, the index $j$, referring to the component under consideration. The method assumes that we have a parameterized form of our estimate, $\hat{\varphi}(y, w)$, where $w$ is a vector of parameters (Taleb and Jutten proposed using a multilayer perceptron with one input and one output for implementing $\hat{\varphi}$; then, $w$ would be the vector of weights of the perceptron). The method uses a minimum mean squared error criterion, defining the mean squared error as

$$
\epsilon = \frac{1}{2}\mathrm{E}\left\{[\hat{\varphi}(Y, w) - \varphi(Y)]^2\right\},
\tag{2.12}
$$

where $\mathrm{E}(\cdot)$ denotes statistical expectation, and the factor $1/2$ is for later convenience. The minimization of $\epsilon$ is performed through gradient descent. The gradient of $\epsilon$ relative to the parameter vector is

$$
\begin{aligned}
\frac{\partial \epsilon}{\partial w} &= \mathrm{E}\left\{[\hat{\varphi}(Y, w) - \varphi(Y)]\frac{\partial \hat{\varphi}(Y, w)}{\partial w}\right\} \\
&= \int p(y)\left[\hat{\varphi}(y, w) - \frac{p'(y)}{p(y)}\right]\frac{\partial \hat{\varphi}(y, w)}{\partial w}\,\mathrm{d}y \\
&= \int p(y)\hat{\varphi}(y, w)\frac{\partial \hat{\varphi}(y, w)}{\partial w}\,\mathrm{d}y - \int p'(y)\frac{\partial \hat{\varphi}(y, w)}{\partial w}\,\mathrm{d}y.
\end{aligned}
$$

The first integral in the last equation is an expected value and the second integral can be computed by integrating by parts, leading to

$$\frac{\partial \epsilon}{\partial \boldsymbol{w}} = \mathrm{E}\left[\hat{\varphi}(y, \boldsymbol{w}) \frac{\partial \hat{\varphi}(y, \boldsymbol{w})}{\partial \boldsymbol{w}}\right]$$
$$- \left[p(y) \frac{\partial \hat{\varphi}(y, \boldsymbol{w})}{\partial \boldsymbol{w}}\right]_{-\infty}^{+\infty} + \int p(y) \frac{\partial^2 \hat{\varphi}(y, \boldsymbol{w})}{\partial y \partial \boldsymbol{w}} \, \mathrm{d}y.$$

The term inside the square bracket in the second line will vanish at infinity, for "well-behaved" distributions, if $\partial \hat{\varphi}(y, \boldsymbol{w})/\partial \boldsymbol{w}$ is limited as a function of $y$, because $p$ will vanish at infinity. When the iterative optimization is close to convergence we shall have $\hat{\varphi} \approx p'/p$ and therefore the term inside that square bracket will be approximately equal to $p'(y)$, which again vanishes at infinity for "well-behaved" distributions, even if $\hat{\varphi}$ approaches an unlimited function. Therefore the square bracket term will vanish for well-behaved distributions, and we are led to the final result

$$\frac{\partial \epsilon}{\partial \boldsymbol{w}} = \mathrm{E}\left[\hat{\varphi}(y, \boldsymbol{w}) \frac{\partial \hat{\varphi}(y, \boldsymbol{w})}{\partial \boldsymbol{w}} + \frac{\partial^2 \hat{\varphi}(y, \boldsymbol{w})}{\partial y \partial \boldsymbol{w}}\right]. \qquad (2.13)$$

This equation allows us to optimize the estimated score function $\hat{\varphi}(y, \boldsymbol{w})$ by any gradient-based method. We use the mean in the training set as an estimate of the expected value that appears in the equation. Note that there is no need to explicitly estimate the densities $p(y_i)$.

INFOMAX, with this method for estimating the score functions, is an iterative method in which two different objective functions are being optimized simultaneously: the separating matrix $W$ is estimated through the maximization of the output entropy, while the score functions are estimated through the minimization of the squared errors in (2.12). An iterative optimization in which there are two or more simultaneous optimization criteria raises some issues:

- What is the best way to interleave optimization iterations for each of the objective functions? A simple solution is to alternate a step of estimation of the unmixing matrix with one of estimation of the score functions. However, it could make more sense to perform several successive steps of optimization of the score functions for each step of optimization of the matrix, so that the optimization of the matrix would be done with rather good estimates of the score functions. Another possibility that has been proposed [97] is to use a stochastic optimization for the score function estimates, in which the expectation is dropped from Eq. (2.13), and the resulting stochastic gradient is used to update the score function estimate once for each pattern that is processed during the optimization.

- More important from a theoretical viewpoint is that this kind of inter leaved optimization of two or more objective functions does not normally have a guaranty of convergence. When a single objective function is being optimized, it is possible to design an optimization algorithm (see [4], for example) that guarantees that the objective function will always improve at each iteration, so that the method will be guaranteed to converge. However, with two interleaved optimizations, an iteration step for one of the objective functions may worsen the value of the other objective function, and vice versa. The consequence is that the process may not converge, possibly having an oscillating or chaotic behavior. This is a theoretical limitation that appears not to have too severe consequences in practice, in the use of this method.

## 2.3  EXPLOITING THE TIME-DOMAIN STRUCTURE

In an ICA setting, if two extracted components are independent, their correlation must be zero.[9] Therefore the correlation matrix of the extracted components $Y_i$ must be diagonal, if these components are independent. This is a necessary condition for independence, but it is not sufficient. There is an infinite number of matrices $W$ that yield uncorrelated components, but only some of them (corresponding to the various permutations of the sources, with all possible scalings) yield statistically independent components.

Let us now assume that the source and mixture vectors are functions of time, denoting them by $S(t)$ and $X(t)$ respectively. If the sources $S_i(t)$ are stochastic processes that are independent from one another, then $S_i(t)$ must be uncorrelated with $S_j(t - \tau)$, if $i \neq j$, for any choice of the delay $\tau$. The separation methods that rely on the temporal structure of the sources enforce the simultaneous decorrelation of $Y_i(t - \tau_i)$ for various sets of delays $\tau_i$, thus increasing the number of constraints that are imposed on the solution. If the sources have temporal correlation, it is normally possible to find sets of delays that force the problem to have only the solutions that correspond to independent components. Linear ICA based on this principle was first proposed by Molgedey and Schuster [75]. Other ICA methods based on the same principle are SOBI (second-order blind identification) [20] and TDSEP (temporal decorrelation source separation) [117]. We shall now briefly analyze them.

Let us define the correlation matrix at time $t$, with delay $\tau$, as

$$C_\tau(t) = \mathrm{E}[Y(t)Y^{\mathrm{T}}(t + \tau)],$$

where $\mathrm{E}(\cdot)$ denotes statistical expectation. If the components of $Y$ are independent, then $C_\tau(t)$ will be diagonal for all values of $t$ and $\tau$. The methods that we are studying implicitly assume that the sources are jointly ergodic. This implies that they are jointly stationary, and therefore

---

[9]Actually it is their covariance that must be zero. But it is common to assume that the means of the random variables have been subtracted out, and then the correlation and the covariance are equal. In the remainder of this chapter we shall assume that the means have been subtracted out.

that $C_\tau(t)$ does not depend on $t$ and can be simply denoted $C_\tau$. More importantly, ergodicity allows us to replace the statistical expectation with a time-domain average, resulting in

$$C_\tau = \langle Y(t)Y^T(t+\tau)\rangle,$$

where $\langle \cdot \rangle$ denotes an average over the time variable $t$. If the components are independent, the $C_\tau$ matrices should be diagonal for all $\tau$. The ICA methods that use the time domain structure enforce that $C_\tau$ be diagonal for more than one value of $\tau$, and differ from one another in the details of how they do so. In practice, $C_\tau$ is replaced with an approximation, obtained by computing an average over a finite amount of time.

The method of Molgedey and Schuster enforces the simultaneous diagonality of $C_0$ (i.e. of the correlation with no delay) and of $C_\tau$ for one specific, user-chosen value of the delay $\tau$. These authors showed that this simultaneous diagonality condition can be transformed into a generalized eigenvalue problem, which can be solved by standard linear algebra methods, yielding the separating matrix $W$. A sufficient condition for separation is

$$R_i(0)R_j(\tau) \neq R_i(\tau)R_j(0) \quad \text{for all} \quad i \neq j, \qquad (2.14)$$

where $R_i(\cdot)$ is the autocorrelation function of the $i$th source.

While this method allows perfect separation in ideal situations, it is some what fragile. A poor choice of the delay $\tau$ can lead to a badly conditioned system (a system in which one or more of the inequalities in (2.14) are actually approximate equalities). This will result in poor separation. The method does not provide any systematic way to choose the delay. SOBI and TDSEP address this difficulty by simultaneously taking into account several delays, so that this problem is alleviated.

SOBI and TDSEP are essentially identical to each other, although they appear to have been independently developed. They start by performing a *prewhitening* (also called *sphering*) operation on the mixture data, i.e. by performing a linear transformation

$$Z = BX$$

such that the zero-delay correlation of $Z$ is the identity matrix

$$E[Z(t)Z^T(t)] = I.$$

Matrix $B$ can be found by one of several methods, one of which is PCA. Once a prewhitening matrix is found, it can be shown that the separating matrix is related to it by

$$W = QB,$$

where $Q$ is an orthogonal matrix [55]. An orthogonal matrix corresponds to a rotation, and possibly a reflection. This means that once the data are prewhitened, ICA can be achieved by a suitable rotation.[10] Because of this, prewhitening is an operation performed by several linear ICA methods, besides SOBI and TDSEP [55].

Once the data are prewhitened, the linear transformations $Y = QZ$ are restricted to rotations, and $C_0$ (computed now using the prewhitened data $Z$ instead of $Y$) is guaranteed to be diagonal (and in fact equal to the identity matrix). SOBI and TDSEP then proceed by considering a set of correlation matrices $C_{\tau_i}$ for a user-chosen set of delay values $\tau_i$ ($i = 1, \ldots, K$), and defining a cost function which is the sum of the squares of the off-diagonal elements of all these matrices.

An approximate joint diagonalization of these matrices is performed by minimizing that cost function by means of a sequence of Givens rotations involving different pairs of coordinates.[11] The simultaneous use, in the cost function, of correlation matrices corresponding to several delays (up to 50 delays in [117]) makes these methods much more robust than the original method of Molgedey and Schuster.

The derivation of these methods does not involve any assumption on the probability distributions of the sources. Therefore they can separate sources with any probability distributions. Namely, they can separate mixtures in which two or more sources have Gaussian distributions. On the other hand, these methods can only separate sources that have temporal or spatial structure. Most sources of practical interest, such as speech, images, or biomedical signals, do have such structure.

SOBI and TDSEP are based only on correlations, which normally are not too sensitive to outliers. Therefore they normally are more robust than methods that rely on higher order statistics such as cumulants.

### 2.3.1    An Example

We show an example of the separation, by TDSEP, of a mixture of two sources: a speech and a music signal. The source signals are shown in Fig. 2.7(a).

These signals were mixed using the mixture matrix

$$A = \begin{bmatrix} 1.0000 & 0.8515 \\ 0.5525 & 1.0000 \end{bmatrix}.$$

The components of the mixture are shown in Fig. 2.7(b). TDSEP was applied to this mixture, using the set of delays $\{0, 1, 2, 3\}$. The separation results are shown in Fig. 2.7(c). They are

---

[10]The reflection is never needed in practice, because it would just change the signs of some components, and the scale indeterminacy of ICA already includes a sign indeterminacy.

[11]Givens rotations of a multidimensional space are rotations that involve only two coordinates at a time, leaving the remaining coordinates unchanged.

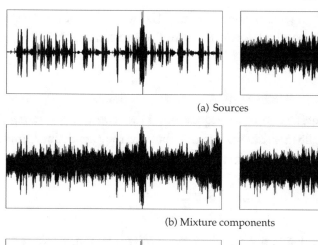

(a) Sources

(b) Mixture components

(c) Separated sources

FIGURE 2.7: Example of source separation using TDSEP

virtually identical to the original sources. This can be confirmed by computing the product of the separating and mixture matrices,

$$WA = \begin{bmatrix} 10.2127 & -0.0288 \\ -0.0286 & 5.9523 \end{bmatrix},$$

which is very close to diagonal.

## 2.4  OTHER METHODS: JADE AND FASTICA

Although we do not intend to provide a complete overview of linear ICA in this chapter, there are some methods that are used so frequently (sometimes as a part of nonlinear separation methods) that they should have a brief mention in this book.

As has been noted above, linear ICA is a rather constrained problem. As such, it can be addressed using optimization criteria that do not fully enforce independence. The two methods that we shall briefly overview in this section, JADE and FastICA, have in common the characteristic of using independence criteria that are only valid within the framework of linear ICA. They also have in common the use of a preprocessing step in which the data are prewhitened, so that only an orthogonal separating matrix (i.e. a rotation of the data) needs to be estimated. This prewhitening step has already been studied in Section 2.3.

## 2.4.1  JADE

Some criteria that are frequently used in linear ICA are based on cumulants. The cumulants of a distribution are related to the coefficients of the expansion of the *cumulant-generating function* as a power series. More specifically, for a univariate distribution, they are defined implicitly as the $\kappa_i$ coefficients in the power series expansion

$$\psi(\omega) = \sum_{i=0}^{\infty} \kappa_i \frac{(j\omega)^i}{i!},$$

where $j$ is the imaginary unit and $\psi$ is the cumulant-generating function, which is defined as the logarithm of the moment-generating function: $\psi(\omega) = \log \varphi(\omega)$. The *moment-generating function* is the Fourier transform of the probability density function,

$$\varphi(\omega) = \int p(x)\, e^{j\omega x}\, dx.$$

The first four cumulants correspond to relatively well-known statistical quantities. The first cumulant of a distribution is its mean, and the second is its variance. The third cumulant is the so-called *skewness*, which is an indicator of the distribution's asymmetry. The fourth cumulant is called *kurtosis*, and is an indicator of the distribution's deviation from Gaussianity. For more details see [55], for example.

Cumulants can also be defined, in a similar way, for joint distributions of several variables. For example, for two random variables $Y_1$ and $Y_2$, the joint cumulants $\kappa_{kl}$, of orders $k$ and $l$ relative to $Y_1$ and to $Y_2$ respectively, are given implicitly by the series expansion

$$\psi(\omega_1, \omega_2) = \sum_{k,l=0}^{\infty} \kappa_{kl} \frac{(j\omega_1)^k}{k!} \frac{(j\omega_2)^l}{l!},$$

where $\psi(\omega_1, \omega_2)$ is the joint cumulant-generating function of $Y_1$ and $Y_2$. Cumulants have several interesting properties, one of which is that, for any two or more independent random variables, the joint cumulants of any order are zero.

Fourth-order cumulants are often used for ICA. Consider a set of jointly distributed random variables $Y_i$. Denote by $\mathrm{cum}(Y_k, Y_l, Y_m, Y_n)$ the coefficient of $(j\omega_k)(j\omega_l)(j\omega_m)(j\omega_n)$ in the expansion of the cumulant-generating function. Some or all of $k, l, m, n$ may be equal to one another. The property that we mentioned above implies that, if the different variables $Y_i$ are independent from one another, this cumulant can be nonzero only if all four arguments are the same, i.e. if $k = l = m = n$, so that a single random variable is involved in the cumulant.

If two or more variables are involved the cumulant must be zero, because those variables are independent from one another.

The four-index array $\text{cum}(Y_k, Y_l, Y_m, Y_n)$ is often called a tensor, and the elements with $k = l = m = n$ are its diagonal elements. Finding a linear transformation $Y = WX$ such that the components of $Y$ are independent can therefore be done by diagonalizing the fourth-order cumulant tensor $\text{cum}(Y_k, Y_l, Y_m, Y_n)$. A widely used algorithm (JADE [26]) uses the diagonalization of a part of this tensor to perform linear ICA.

### 2.4.2   FastICA

Another criterion that is used for linear ICA is based on *negentropy*, a concept that we shall define next. Recall that, among all distributions with a given variance, the Gaussian distribution is the one with the largest entropy (see Appendix A.2). Consider a random variable $Y$, and let $Z$ be a Gaussian random variable with the same variance. The quantity

$$J(Y) = H(Z) - H(Y)$$

is nonnegative, and is zero only if $Y$ is Gaussian. Therefore $J(Y)$, called the negentropy of $Y$, is a measure of the deviation of $Y$ from Gaussianity.

Mixtures of independent random variables tend to have distributions that are closer to Gaussian than the original random variables (an expression of this is the central limit theorem). Consequently, once the mixture data have been prewhitened, finding a rotation that yields components that are maximally non-Gaussian is a way to identify the original unmixed random variables, and therefore to perform ICA. Negentropy can be used as the measure of non-Gaussianity.

Negentropy is, however, somewhat hard to use, because it needs an estimation of the entropy and therefore of the pdf. FastICA works by computing an efficient approximation of negentropy, and using an iteration that quickly converges to the rotation that yields its maximum. It can be used for extracting one independent component at a time, or for simultaneously extracting a set of independent components. For more details see [54].

## 2.5   SUMMARY

We have made a short overview of linear source separation and linear ICA, giving a brief introduction to some of the methods that are used to perform them. Among these we can distinguish two classes. Methods such as INFOMAX, JADE, and FastICA are based (explicitly or implicitly) on statistics of order higher than 2, and demand that at most one source be Gaussian. This means that their performance will be poor if more than one source is close to Gaussian. However, most of the sources that are of practical interest have distributions that deviate markedly from Gaussian (for example, speech is strongly supergaussian, while images tend to be strongly subgaussian). Methods of this class do not need (and do not exploit) any temporal or spatial structure of the sources. On the other hand, methods such as TDSEP and SOBI use only second-order statistics, and can deal with any number of Gaussian sources. However, they demand sources with temporal structure (or spatial structure, as in images). Again, most sources of practical interest (such as speech, biomedical signals, or images) do have temporal or spatial structure that can be used.

In this overview of linear blind source separation it was not our intention to be complete. To keep this work compact, we limited ourselves to a short introduction to the linear ICA/BSS problem and to a reference to the main methods that are of interest to the study of nonlinear separation, which is the main subject of this book.

CHAPTER 3

# Nonlinear Separation

Nonlinear source separation, as discussed in this book, deals with the following problem: There is an unobserved random vector $S$ whose components are called sources and are mutually statistically independent. We observe a mixture vector $X$, which is related to $S$ by

$$X = \mathcal{M}(S),$$

$\mathcal{M}$ being a vector-valued invertible function, which is, in general, nonlinear. Unless stated otherwise, the size of $X$ is assumed to be equal to the size of $S$; i.e., we consider a so-called square problem. We wish to find a transformation

$$Y = \mathcal{F}(X),$$

where $\mathcal{F}$ is a vector-valued function and the size of $Y$ is the same as those of $X$ and $S$, such that the components of $Y$ are equal to the components of $S$, up to some indeterminacies to be clarified later. These indeterminacies will include, at least, permutation and scaling, as in the linear case. As we can see, we are considering only the simplest nonlinear setting, in which there is no noise and the mixing is both instantaneous and invariant.

There is a crucial difference between linear and nonlinear separation, which we should immediately emphasize. While in a linear setting, the independence of the components of $Y$ suffices, under rather general conditions, to guarantee the recovery of the sources (up to permutation and scaling), there is no such guaranty in the nonlinear setting. This has been shown by several authors [30, 55, 73]. We shall show it for the case of two sources, since the generalization to more sources is then straightforward. Assume that we have a two-dimensional random vector $X$ and that we form the first component of the output vector $Y$ as some function of the mixture components,

$$Y_1 = f_1(X_1, X_2).$$

The function $f_1$ is arbitrary, subject only to the conditions of not being constant and of $Y_1$ having a well-defined statistical distribution. Now define an auxiliary random variable $Z$ as

$$Z = f_2(X_1, X_2),$$

$f_2$ being arbitrary, subject only to the conditions of the variables $Y_1$ and $Z$ having a well-defined joint distribution and of $F_{Z \mid Y_1}(z \mid y_1)$ being continuous in $z$. Define the second component of $\boldsymbol{Y}$ as

$$Y_2 = F_{Z \mid Y_1}(Z \mid Y_1).$$

For each value of $Y_1$, $Y_2$ is the CDF of $Z$ given that value of $Y_1$. Therefore $Y_2$ will be uniformly distributed in $(0, 1)$ for all values of $Y_1$ (see Appendix A.1). As a consequence, $Y_2$ will be independent from $Y_1$, and $\boldsymbol{Y}$ will have independent components.

Given the large arbitrariness in the choice of both $f_1$ and $f_2$, we must conclude that there is a very large variety of ways to obtain independent components. This variety is much larger than the variety of solutions in which the sources are separated. Therefore, in most cases, each of the components of $\boldsymbol{Y}$ will depend on both $S_1$ and $S_2$. This means that the sources will not be separated at the output, even though the output components will be independent.

This raises an important difficulty: if we perform nonlinear ICA, i.e. if we extract independent components from $\boldsymbol{X}$, we have no guaranty that each component will depend on a single source. ICA becomes ill-posed, and ICA alone is not a means for performing source separation, in the nonlinear setting.

Researchers have addressed this difficulty in three main ways:

- One way has been to restrict the range of allowed nonlinearities, so that ICA becomes well-posed again. When we restrict the range of allowed nonlinearities, an important consideration is whether there are practical situations in which the nonlinearities are (at least approximately) within our restricted range. The case that has met a significant practical applicability is the restriction to *post-nonlinear* (PNL) mixtures, described ahead.

- Another way to address the ill-posedness of nonlinear ICA has been the use of regularization. This corresponds to placing some "soft" extra assumptions on the mixture process and/or on the source signals, and trying to smoothly enforce those assumptions in the separation process. The kind of assumption that has most often been used is that the mixture is only mildly nonlinear. This is the approach taken in the MISEP and ensemble learning methods, to be studied in Sections 3.2.1 and 3.2.2 respectively.

- The third way to eliminate the ill-posedness of nonlinear ICA has been to use extra structure (temporal or spatial) that may exist in the sources. Most sources of practical interest have temporal or spatial structure that can be used for this purpose. This is the approach taken in the kTDSEP method, to be studied in Section 3.2.3. This approach is also used, together with regularization, in the second application example of the ensemble learning method (p. 62).

In the next sections we shall study, in some detail, methods that use each of these approaches, starting with the constraint to PNL mixtures.

## 3.1 POST-NONLINEAR MIXTURES

There are several sets of constraints that can be imposed on the nonlinear ICA problem in order to eliminate its ill-posedness, making the solution essentially unique. However, such constraints are of interest only if they are known to correspond (at least approximately) to practical situations. Among the constraints that have been proposed, the constraint to the so-called *PNL mixtures* is, to date, the only one that has been found to have a significant practical applicability. In this section we shall study the separation of PNL mixtures in some detail.

Consider the mixture process depicted in Fig. 3.1. The source signals are first subject to a linear mixture, and then each mixture component suffers a nonlinear, invertible transformation. Formally, we can express this mixture process as follows. There is a set of independent sources $S_i$, forming the source vector $S$. An intermediate mixture vector $U$ is formed by a linear mixture process

$$U = AS.$$

The observed mixture $X$ is obtained through component-wise nonlinearities applied to the components of the linear mixture,

$$X_i = f_i(U_i),$$

where the functions $f_i$ are invertible. As before, we shall assume that the sizes of $S$, $U$, and $X$ are the same.

This is the structure of the so-called *PNL* mixture process. Its interest resides mainly in the fact that it corresponds to a well-identified practical situation. PNL mixtures arise whenever, after a linear mixing process, the signals are acquired by sensors that are nonlinear. This is a situation that sometimes occurs in practice. The other fact that makes these mixtures interesting is that they present almost the same indeterminations as linear mixtures, as we shall see below.

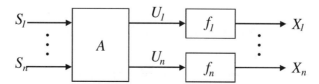

FIGURE 3.1: Post-nonlinear mixture process. Block $A$ performs a linear mixture (i.e. a product by a matrix). The $f_i$ blocks implement nonlinear invertible functions

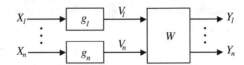

**FIGURE 3.2:** Separating structure for post-nonlinear mixtures. The $g_i$ functions should, desirably, be the inverses of the corresponding $f_i$ in the mixture process. Block $W$ is linear, performing a product by a matrix

Naturally, the separation of PNL mixtures uses a structure that is a "mirror" of the mixing structure (Fig. 3.2). Each mixture component is first linearized by going through a nonlinearity, whose purpose is to invert the nonlinearity present in the mixture process, and then the linearized mixture goes through a standard linear separation stage. Formally,

$$V_i = g_i(X_i)$$
$$\boldsymbol{Y} = W\boldsymbol{V}.$$

The indeterminations that exist in PNL separation are only slightly wider than those of linear ICA. It has been shown [97] that, under a set of conditions indicated ahead, if the components of $\boldsymbol{Y}$ are independent, they will obey

$$\boldsymbol{Y} = PD\boldsymbol{S} + \boldsymbol{t},$$

in which $P$ is a permutation matrix, $D$ is a diagonal matrix, and $\boldsymbol{t}$ is a vector. This means that the sources are recovered up to an unknown permutation (represented by $P$), unknown scalings (represented by $D$), and unknown translations (represented by $\boldsymbol{t}$).

The unknown translations represent an additional indetermination relative to the linear case. However, this indetermination is often not too serious a problem because most sources can be translated back to their original levels after separation, using prior knowledge. For example, speech and other acoustic signals normally have a mean of zero, and in images we can often assume that the lowest intensity corresponds to black.

The aforementioned result is valid if

- the mixture matrix $A$ is invertible and has at least two nonzero elements per row and/or per column;
- the functions $f_i$ are invertible and differentiable; and
- the pdf of each source is zero in one point at least.

The last condition precludes the possibility of having Gaussian sources, but is not too restrictive since most distributions can, in practice, be considered bounded.[12] The fact that the sources can be recovered up to permutation, scaling, and translation implies that each $g_i$ must be the inverse of the corresponding $f_i$, up to unknown translation and scaling, and thus that the nonlinearities of the mixture can be estimated, up to unknown translations and scalings.

The need for the first condition (that $A$ must have at least two nonzero elements per row and/or per column) can be understood in the following way: If $A$ had just one nonzero element per row and per column, it would correspond just to a permutation and a scaling. The $U_i$ components would be independent from one another, and so would be the $X_i$, for any nonlinearities $f_i$. Therefore it would be impossible to estimate and invert the nonlinearities $f_i$ on the basis of the independence criterion alone. One would only be able to recover the sources up to a permutation and an arbitrary invertible nonlinear transformation per component: the nonlinearities could not be compensated for. For it to be possible to estimate and invert the nonlinearities there must be some mixing in $A$.

PNL mixtures have been subject to extensive study, e.g. [1,2,15,95–97]. We shall review the basic separation method.

### 3.1.1   Separation Method

The basic method for the separation of PNL mixtures appears to have been independently proposed, in 1997, by Taleb and Jutten [95] and by Lee *et al.* [67]. A fuller account is given in [97]. The method is based on the minimization of the mutual information of the extracted components, $I(Y)$. There are several variants of the method, differing mostly in the way in which the inverse nonlinearities $g_i$ are represented and in the way the score functions (see Section 2.2.3) are estimated. The representation of the $g_i$ functions has been made in several ways, including multilayer perceptrons (MLPs), piecewise-linear functions, and polynomials. The score functions have been approximated by means of MLPs, using the method of Section 2.2.3, and have also been computed, through Eq. (2.11), from the pdfs, which can be estimated by any standard method. The important fact to retain at this point is that the method uses representations of both the inverse nonlinearities, $\hat{g}_i(x_i, \boldsymbol{\theta}_i)$, and the score functions, $\hat{\varphi}_i(y_i, \boldsymbol{w}_i)$, where $\boldsymbol{\theta}_i$ and $\boldsymbol{w}_i$ are vectors of parameters.

The minimization of the mutual information $I(Y)$ is made relative to the separating matrix $B$ and to the parameter vectors $\boldsymbol{\theta}_i$, and is made iteratively through gradient descent. A full presentation of the method is given in [97]. Here we shall give only the expression of the gradient from which the parameter update equations can readily be found.

---

[12]This condition has been relaxed in [1] to the condition that the joint density of the sources be continuous and differentiable, with no more than one Gaussian source.

The gradient of the mutual information is given by

$$\frac{\partial I(\boldsymbol{Y})}{\partial W} = -W^{-\mathrm{T}} - \mathrm{E}(\boldsymbol{\varphi} \boldsymbol{V}^{\mathrm{T}}) \tag{3.1}$$

$$\frac{\partial I(\boldsymbol{Y})}{\partial \boldsymbol{\theta}_i} = -\mathrm{E}\left[\frac{\partial \log |g_i'(Y_i, \boldsymbol{\theta}_i)|}{\partial \boldsymbol{\theta}_i}\right]$$

$$-\mathrm{E}\left[\left(\sum_{k=i}^{n} \varphi_k(Y_k) w_{ki}\right) \frac{\partial g_i(Y_i, \boldsymbol{\theta}_i)}{\partial \boldsymbol{\theta}_i}\right].$$

In these equations, we have used the following notational conventions:

- $W^{-\mathrm{T}} = (W^{-1})^{\mathrm{T}}$.

- $\boldsymbol{V}$ is the vector whose components are defined by (3.1).

- $\mathrm{E}(\cdot)$ denotes statistical expectation.

- $g_i'(Y_i, \boldsymbol{\theta}_i) = \frac{\partial g_i(Y_i, \boldsymbol{\theta}_i)}{\partial Y_i}$.

- $w_{ki}$ is the element $ki$ of matrix $W$.

- $\boldsymbol{\varphi}$ is the column vector $[\varphi_1(Y_1), \varphi_2(Y_2), \ldots, \varphi_n(Y_n)]^{\mathrm{T}}$, $n$ being the size of $\boldsymbol{Y}$, i.e. the number of sources and of extracted components.

In practice, the score function estimates $\hat{\varphi}_i(Y_i, \boldsymbol{w}_i)$ are used instead of the actual score functions $\varphi_i(Y_i)$ in these equations. Also, instead of the statistical expectations $\mathrm{E}(\cdot)$, what is used in practice are the means computed in the training set.

This method is, essentially, an extension of INFOMAX to the PNL case. In particular, note that the expression of the gradient relative to the separation matrix, (3.1), is essentially the same as in INFOMAX, (2.7), if we replace the statistical expectation $\mathrm{E}(\cdot)$ with the mean in the training set $\langle \cdot \rangle$.[13] And, as in INFOMAX, people have used the relative/natural gradient method (see Section 2.2) for the estimation of this matrix.

This PNL separation method is also subject to the issues concerning the interleaved optimization of two different objective functions, which we mentioned in Section 2.2.3, since the estimation of the score functions is done through a criterion different from the one used for the estimation of $W$ and of the $\boldsymbol{\theta}_i$. Once again, however, these issues do not seem to raise great difficulties in practice.

Several variants of the basic PNL separation algorithm have appeared in the literature. Some of them have to do with different ways to estimate the source densities (or equivalently,

---

[13]There is a sign difference, because here we use as objective function the mutual information, which is to be minimized, and in INFOMAX we used the output entropy, which was to be maximized. Recall that the two are equivalent—see Eq. (2.2).

the score functions) [94, 97], while other ones have to do with different ways to represent the inverse nonlinearities $g_i$ [2, 96, 97]. An interesting development was the proposal, in [98], of a nonparametric way to represent the nonlinearities,[14] in which each of them is represented by a table of the values of $g_i(x_i)$ for all the elements of the training set (using interpolation to compute intermediate values, if necessary). This appears to be one of the most flexible and most powerful ways to represent these functions.

An efficient way to obtain estimates of the $g_i$ nonlinearities is to use the fact that mixtures tend to have distributions that are close to Gaussians. This idea has been used in [92, 114–116]. The last three of these references used the TDSEP method (cf. Section 2.3), instead of INFOMAX, to perform the linear part of the separation.

Other variants of PNL methods, involving somewhat different concepts, have been proposed in [16, 17, 67]. In addition, a method based on ensemble learning (discussed ahead, in Section 3.2.2) has been proposed in [57]. It can separate mixtures in which some of the individual nonlinearities are not invertible, but in which there are more mixture components than sources. In [60] it was shown that when the sources have temporal structure, incorporating temporal decorrelation into the PNL separation algorithm can achieve better separation. A method using genetic algorithms in the optimization process was proposed in [86]. A quadratic dependence measure has been proposed in [3], and has been used for the separation of PNL mixtures.

### 3.1.1.1 Extensions of the Post-Nonlinear Setting

Some extensions of PNL mixtures have been considered in the literature. The two most important ones concern convolutive PNL mixtures and Wiener systems. Convolutive PNL mixtures assume that the linear part of the mixture is not instantaneous, but instead involves time delays. They have been studied in [14]. Wiener systems consist of a linear filter, operating on an independent, identically distributed (and therefore white) input signal, the filter being followed by an invertible nonlinearity. These systems have been considered in [93, 98].

### 3.1.2 Application Example

We present an example of the application of PNL separation to an artificial mixture of two sources.[15] The sources were the two sinusoids shown in Fig. 3.3(a). The mixture matrix was

$$A = \begin{bmatrix} 1.0 & 0.6 \\ 0.7 & 1.0 \end{bmatrix}.$$

---

[14]"Nonparametric," in this context, does not mean that there are no parameters. It means that there is not a fixed number of parameters, but instead that this number is of the order of magnitude of the number of training data, and grows with it.

[15]This example was worked out by us using software made publicly available by its authors (see Appendix B).

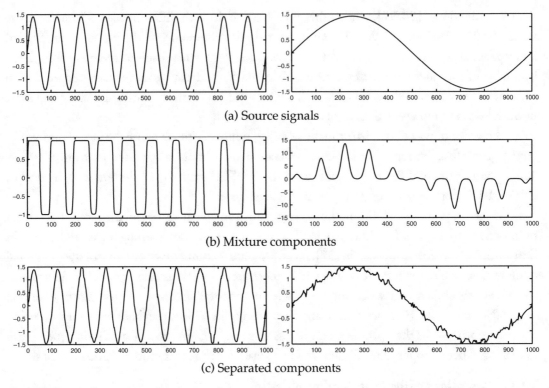

(a) Source signals

(b) Mixture components

(c) Separated components

FIGURE 3.3: Separation of a post-nonlinear mixture: source, mixture, and separated signals

The nonlinearities were

$$f_1(u_1) = \tanh(5u_1)$$
$$f_2(u_2) = (u_2)^3.$$

They are plotted in Fig. 3.4(a). The mixture components are shown in Fig. 3.3(b), where we can see the strong distortions caused by the nonlinearities. The 1000-sample segments shown in the figure were used as training set.

The separation was performed through the algorithm indicated above, with the following specifics:

- The inverse nonlinearities $g_i$ were represented by the nonparametric method proposed in [98] and mentioned above.

- The score functions were estimated through their definition, Eq. (2.11). For use in this equation, the densities $p(y_i)$ were estimated using kernel density estimators with a Gaussian kernel.

- The optimization was performed for 200 epochs.

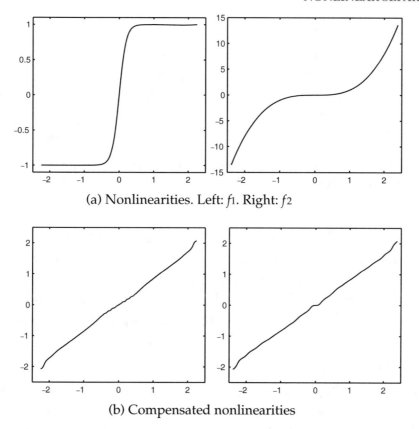

(a) Nonlinearities. Left: $f_1$. Right: $f_2$

(b) Compensated nonlinearities

FIGURE 3.4: Separation of a post-nonlinear mixture: Nonlinearities and their compensation

Fig. 3.3(c) shows the separation results. The source signals were recovered with a relatively small amount of error. Fig. 3.4(b) shows the compensated nonlinearities, i.e. the functions $g_i[f_i(\cdot)]$. We can see that the compensation had a rather good accuracy. The good recovery of the source signals implies that the separating matrix was well estimated. Its product with the mixing matrix was

$$WA = \begin{bmatrix} 1.14 & -0.16 \\ 0.07 & 1.28 \end{bmatrix},$$

which is close to diagonal.

Fig. 3.5 shows the scatter plots corresponding to this example. The mixture scatter plot gives a good idea of the very strong nonlinear distortions that were involved. There are two regions (the two "vertical segments") in which the mixture was almost singular, due to the strong saturation performed by $f_1$. This makes the mixture rather hard to separate. Despite this, the method was able to recover the sources with a reasonably good accuracy.

FIGURE 3.5: Separation of a post-nonlinear mixture: scatter plots. From left to right: sources, mixture components, separated components

## 3.2 UNCONSTRAINED NONLINEAR SEPARATION

In this section we shall present, in three separate subsections, the three main methods that have been proposed for unconstrained nonlinear source separation: MISEP, ensemble learning, and kTDSEP. A fourth subsection then gives an historical overview of the area, briefly mentioning several other methods that have been proposed, but that have had more restricted use.

### 3.2.1 MISEP

MISEP (for Mutual Information-based SEParation) is an extension of the INFOMAX linear ICA method to the nonlinear separation framework, and was proposed by L. Almeida [5, 9]. It is based on the structure of Fig. 3.6. This structure is very similar to the one of INFOMAX (Fig. 2.1), the only difference being that the separation block is now nonlinear. Like INFOMAX, the method uses the minimization of the mutual information of the extracted components, $I(Y)$, as an optimization criterion. Also as in INFOMAX, this minimization is transformed into the maximization of the output entropy $H(Z)$ through the reasoning of Eqs. (2.2), assuming again that each $\psi_i$ function equals the CDF of the corresponding $Y_i$ variable.

For implementing the nonlinear separation block $\mathcal{F}$, the method can use essentially any nonlinear parameterized system. In practice, the kind of system that has been used most

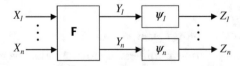

FIGURE 3.6: Structure used by MISEP. $\mathcal{F}$ is a generic nonlinear block, and is what performs the separation proper. The separated outputs are $Y_i$. The $\psi_i$ blocks, implementing nonlinear increasing functions, are auxiliary, being used only during optimization

frequently is an MLP.[16] For this reason, we shall present the method using as example a structure based on an MLP. To make the presentation simple, we shall use a very simple MLP structure with a single hidden layer of nonlinear units and a layer of linear output units, and with connections only between successive layers. We note, however, that more complex structures can be used, and have actually been used in the examples that we shall present ahead.

MISEP uses the maximum entropy method for estimation of the $\psi_i$ functions, as explained in Section 2.2.2. And in fact, as we mentioned in that section, once that estimation method is understood, the extension to nonlinear MISEP is quite straightforward. In what follows we shall assume that the contents of that section have been well understood. If the reader has only glossed over that section, we recommend reading it carefully now, before proceeding with nonlinear MISEP.

### 3.2.1.1 Maximizing the Output Entropy

Assuming that the $\mathcal{F}$ block is implemented by means of an MLP, as said above, and that the $\psi_i$ blocks are also implemented through MLPs (as in Section 2.2.2), the whole system of Fig. 3.6 can be seen as a single, global MLP. We wish to optimize it through maximization of the output entropy. Following the reasoning presented in Section 2.2.2, we repeat Eq. (2.10):

$$\mathcal{L} = \log \left| \det J \right|.$$

Recall that $\mathcal{L}$ is one of the additive terms forming the objective function and is relative to one of the training patterns (for simplicity we have again dropped the subscript referring to the training patterns). Also recall that $J = \partial z / \partial x$ is the Jacobian of the transformation made by the system.

We need to compute the gradient of $\mathcal{L}$ relative to the network's weights to be able to use gradient-based optimization methods. As in Section 2.2.2, we shall compute that gradient through backpropagation. We shall use a network that computes $J$ and shall backpropagate through it to find the desired gradient.

**A Network That Computes the Jacobian**   The network that computes $J$ is constructed using the same reasoning that was used in Section 2.2.2. In fact, the MLP that implements the system of Fig. 3.6 is very similar to the one that implements the system of Fig. 2.1. The only difference is that the linear block $W$ of Fig. 2.1 is now replaced with an MLP, which has a hidden layer of nonlinear units followed by a layer of linear output units.

---

[16]In [7] a system based on radial basis functions was used. In [11] a system using an MLP together with product units was used.

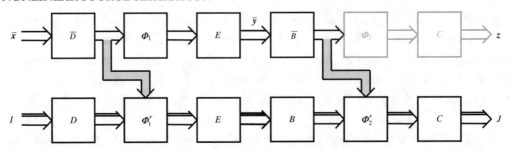

**FIGURE 3.7:** Network for computing the Jacobian. The lower part is what computes the Jacobian proper and is essentially a linearized version of the network of Fig. 3.6. The upper part is identical with the network of Fig. 3.6, but is drawn in a different way. The two upper-right blocks, shown in gray, are not necessary for computing $J$ and are shown for reference only. See the text for further explanation

We have already seen, in Section 2.2.2, how to deal with a layer of nonlinear hidden units in constructing the network that computes the Jacobian. The structure of the network, for the current case, is shown in Fig. 3.7. In this figure, the lower part is what computes the Jacobian proper. The upper part is identical to the network of Fig. 3.6, but is drawn in a different way. We shall start by describing this part.

- The input vector $\bar{x}$ is the input pattern $x$, augmented with a component equal to 1, which is useful for implementing the bias terms of the hidden units of block $\mathcal{F}$.
- Blocks $\bar{D}$, $\Phi_1$, and $E$ form block $\mathcal{F}$:
  - Block $\bar{D}$ implements the product by the weight matrix (which we also call $\bar{D}$) that connects the input to the hidden layer. This matrix includes a row of elements corresponding to the bias terms (this is indicated by the overbar in the matrix's symbol). The output of this block is formed by the vector of input activations of the hidden units of $\mathcal{F}$.
  - $\Phi_1$ is a nonlinear block that applies to this vector of input activations, on a per-component basis, the nonlinear activation functions of the hidden units of $\mathcal{F}$. This block's output is the vector of output activations of that hidden layer.
  - Block $E$ implements the product by the weight matrix (also designated by $E$) connecting the hidden layer to the output layer of $\mathcal{F}$. The block's output is the vector of extracted components, $y$. In the figure it is shown as $\bar{y}$ because it is considered to be augmented with a component equal to 1, which is useful for implementing the bias terms of the next layer of nonlinear units.

- As before, in Section 2.2.2, the three rightmost blocks represent all the $\psi_i$ blocks of Fig. 3.6 taken together:

  - Block $\bar{B}$ performs a product by matrix $\bar{B}$, which is the weight matrix of the hidden units of all the $\psi_i$ blocks, taken as forming a single hidden layer. The overbar in $\bar{B}$ indicates that this matrix incorporates elements relative to the bias terms of the hidden units. The output of this block consists of the vector of input activations of the units of the hidden layer of all the $\psi_i$ blocks taken together.

  - Block $\Phi_2$ is nonlinear and applies to this vector of input activations, on a per-component basis, the nonlinear activation functions of the hidden units. The output of this block is the vector of output activations of the global hidden layer of the $\psi_i$ blocks.

  - Block $C$ performs a product by matrix $C$, which is the weight matrix of the linear output units of the $\psi_i$ blocks. The output of the $C$ block is the output vector $\boldsymbol{z}$.

As before, and since there are no interconnections between the various $\psi_i$ blocks, matrices $\bar{B}$ and $C$ have a special structure, with a large number of elements equal to zero. This does not affect our further analysis, except for the fact that during the optimization these elements are not updated, being always kept equal to zero.

We shall now describe the lower part of the network, which is the part that computes the Jacobian proper. The Jacobian is given by

$$J = C\Phi_2' BE\Phi_1' D, \tag{3.2}$$

in which

- Matrices $D$, $E$, $B$, and $C$ are as defined above, the absence of an overbar indicating that the matrices have been stripped of the elements relating to bias terms (which disappear when differentiating relative to $\boldsymbol{x}$).

- Block $\Phi_1'$ performs a product by a diagonal matrix whose diagonal elements are the derivatives of the activation functions of block $\Phi_1$, for the input activations that these activation functions receive for the current input pattern.

- Block $\Phi_2'$ is similar to $\Phi_1'$ but performs a product by the derivatives of the activation functions of block $\Phi_2$.

To compute the values of the activation functions' derivatives, blocks $\Phi_1'$ and $\Phi_2'$ need to receive the input activations of the corresponding units. This information is supplied by the upper part of the network through the gray-shaded arrows. The reason for the presence of the upper part of the network is exactly supplying this information.

### 3.2.1.2 Optimization Procedure

The optimization procedure is essentially identical to what was described in Section 2.2.2. There is one important additional aspect that we shall discuss, however, which concerns optimization speed. We start by recalling the main points from Section 2.2.2, and then discuss the optimization speed issue. The main aspects to recall are as follows:

- The input to the backpropagation network is

$$\frac{\partial \mathcal{L}}{\partial J} = J^{-\mathrm{T}}. \qquad (3.3)$$

- The backpropagation through the $\Phi'$ blocks is performed as discussed in Section 2.2.2 and depicted in Fig. 2.4.

- The gradient computation must take into account the shared weights that exist in the network.

- The constraints on the $\psi_i$ networks (initialization with nonnegative weights and renormalization of the output weights) must be implemented.

The procedure that we have described allows us to compute the gradient of the objective function and therefore permits the use of any gradient-based optimization method. Among these, the simplest is plain gradient descent, possibly augmented with the use of momentum for better performance [4, 39]. While this method, with suitably chosen step size and momentum coefficients, will normally be fast enough to perform linear separation, as described in Section 2.2.2, nonlinear separation usually leads to much more complex optimization problems, which can only be efficiently handled by means of accelerated optimization methods.

To obtain good results in a reasonable amount of time it is therefore essential to use a fast optimization procedure. Among these, we mention the use of gradient descent with adaptive step sizes (see [4], Sections C.1.2.4.2 and C.1.2.4.3). This method is simple to implement and has been used with MISEP with great success. Another important set of fast optimization methods that are good candidates for use with MISEP (but which have not been used yet, to our knowledge) is the class of conjugate gradient methods [39].

### 3.2.1.3 Network Structure

After having understood the MISEP method, the reader may ask why we need to use the special network structure depicted in Fig. 3.6. Given that the separator block $\mathcal{F}$ is already nonlinear, it would seem that we would not need the $\psi_i$ blocks. We could simply maximize the entropy at the output of $\mathcal{F}$, restricting this output to be within a hypercube, and we would obtain a distribution that would be approximately uniform within that hypercube. Therefore its components $Y_i$ would be independent from one another.

This reasoning is correct, but does not take into account that to deal with the indetermination of nonlinear ICA, we need to assume that the mixture is mildly nonlinear, and we need to ensure that the separator also performs a mildly nonlinear transformation. In the scenario presented in the last paragraph, the extracted components $Y_i$ would have approximately uniform distributions. If the sources that were being handled did not have close-to-uniform distributions (and many real-life sources are not close to uniform), the $\mathcal{F}$ block would have to perform a strongly nonlinear transformation to make them uniform, and it would not be possible to keep $\mathcal{F}$ mildly nonlinear. With the structure of Fig. 3.6, we can keep $\mathcal{F}$ mildly nonlinear, while allowing the $\psi_i$ to be strongly nonlinear, so that the $Z_i$ components become close to uniform. It is even possible, if desired, to apply explicit regularization to $\mathcal{F}$ to force it to stay close to linear. In the application examples presented ahead, regularization was applied to $\mathcal{F}$ to ensure a correct separation of the sources.

### 3.2.1.4 Application Examples

We shall now present some examples of the application of the MISEP method, first to artificial mixtures and then to a real-life situation.

**Artificial Mixtures** We show two examples of mixtures of two sources. In the first case both sources were supergaussian, and in the second case one was supergaussian and the other subgaussian (and bimodal). Fig. 3.8(a) shows the joint distributions of the sources for the two cases.

The nonlinear part of the mixtures was of the form

$$\hat{x}_1 = s_1 + a(s_2)^2$$
$$\hat{x}_2 = s_2 + a(s_1)^2,$$

with a suitably chosen value of $a$. The mixture components were then obtained by rotating the vector $\hat{x}$ by $45°$:

$$x = A\hat{x},$$

with

$$A = \frac{1}{\sqrt{2}} \begin{bmatrix} 1 & 1 \\ -1 & 1 \end{bmatrix}.$$

Fig. 3.8(b) shows the mixture distributions.

Fig. 3.8(c) shows the results of linear separation. As expected, linear ICA was not able to undo the nonlinear part of the mixtures and just performed rotations of the mixture distributions.

For nonlinear separation, block $\mathcal{F}$ was formed by an MLP with a hidden layer with 40 sigmoidal units and with linear output units. Of the 40 hidden units, 20 were connected only to one of the block's output units and another 20 to the other output unit. Direct "shortcut" connections between the input and output units of block $\mathcal{F}$ were also implemented, so that the network could exactly perform a linear separation, by setting all the weights of the hidden units to zero.

The $\psi$ blocks were formed by MLPs with a single hidden layer, with 10 sigmoidal hidden units, and with a linear output unit. The training set was formed, in both cases, by 1000 randomly chosen mixture vectors.

Fig. 3.8(d) shows the results of nonlinear separation. We can see that the sources were well recovered in both cases. This was possible, in spite of the ill-posedness of nonlinear ICA, for two reasons:

- The mixtures were not too strongly nonlinear.
- The separation system had inherent regularization, which biased it toward implementing close-to-linear transformations. This regularization was achieved by two means:
  - The $\mathcal{F}$ block was initialized so as to implement purely linear transformations (all output weights of the hidden layer were initialized to zero).
  - MLPs inherently tend to yield relatively smooth nonlinearities, if their nonlinear units' weights are initialized to small values.

In what concerns processing time for nonlinear separation, the first example (the one with two supergaussian sources) took 1000 training epochs to converge, corresponding approximately to 3 min on a 1.6 GHz Pentium-M (Centrino) processor programmed in Matlab. The second example (with a supergaussian and a subgaussian source) took 500 epochs, corresponding to approximately 1.5 min on the same processor. This shows that MISEP is relatively fast, at least for situations similar to those presented in these examples.

Results similar to these were presented in [8] for artificial mixtures of four sources (two supergaussian and two bimodal) and in [9] for mixtures of 10 supergaussian sources. In the latter case it was found that the number of epochs for convergence did not depend significantly on the number of sources and that the computation time per epoch increased approximately linearly with the number of sources for this range of problem dimensionalities. Therefore MISEP remained relatively fast in all these situations.

**Real-Life Mixtures**   We now present results of the application of MISEP to a real-life image separation problem. When we acquire an image of a paper document (using a scanner, for example), the image printed on the back page sometimes shows through due to partial transparency

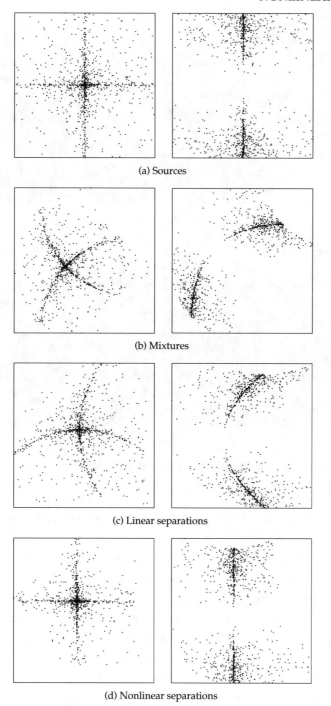

(a) Sources

(b) Mixtures

(c) Linear separations

(d) Nonlinear separations

**FIGURE 3.8:** Separation of artificial mixtures. Left column: two supergaussian sources. Right column: a supergaussian and a bimodal source

(a) Source images

(b) Acquired (mixture) images

FIGURE 3.9: Image separation example: source and mixture images

of the paper. The mixture that is thus obtained is nonlinear. Since it is possible to acquire both sides of the paper, we can obtain a two-component nonlinear mixture, which is a good candidate for nonlinear separation. Here we present a difficult instance of this problem, in which the paper is of the "onion skin" type, yielding a strong, rather nonlinear mixture.

Fig. 3.9(a) shows the two sources. These images were printed on opposite sides of a sheet of onion skin paper. Both sides of the paper were then acquired with a desktop scanner. Fig. 3.9(b) shows the images that were acquired. These images constitute the mixture components. Fig. 3.10(a) shows the joint distribution of the sources, while Fig. 3.10(b) shows the mixture distribution. It is easy to see that the mixture was nonlinear because the source distribution was contained within a square, and a linear mixture would have transformed this square into a parallelogram. It is clear that the outline of the mixture distribution is far from a parallelogram. The distortion of the original square outline gives an idea of the degree of nonlinearity of the mixture.

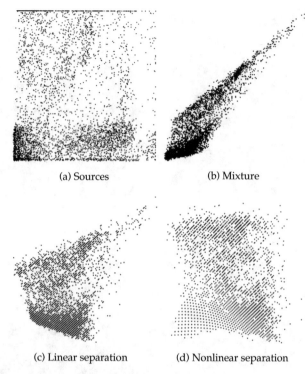

(a) Sources        (b) Mixture

(c) Linear separation       (d) Nonlinear separation

FIGURE 3.10: Scatter plots corresponding to the image separation example

The separation of this mixture was performed both with the linear ICA and with a non-linear, MISEP-based system. Fig. 3.11 shows the results. We can see that nonlinear separation yielded significantly better results than linear separation. The distributions of the separated components are shown in Figs. 3.10(c) and 3.10(d) for the linear and nonlinear cases respectively. These distributions also confirm the better separation achieved by the nonlinear system.

The quality of separation was also assessed by means of several objective quality measures, again showing the advantage of nonlinear separation. One of these measures, the mean signal-to-noise ratio of the separated sources, for a set of 10 separation tests, is shown in Table 3.1. We see that nonlinear separation yielded an improvement of 4.1 dB in one of the sources, and of 3.4 dB in the other one, relative to linear separation.

In these separation tests, the $\mathcal{F}$ block had the same structure as the one used in the previous artificial mixture examples. The $\psi$ blocks had 20 hidden units each, to be able to approximate well the somewhat complex cumulative distributions of the sources. The training set was formed by the intensities from 5000 pixel pairs, randomly chosen from the mixture images. Separation was achieved in 400 training epochs, which took approximately 9 min on a 1.6 GHz Pentium-M (Centrino) processor programmed in Matlab.

(a) Linear separation

(b) Nonlinear separation by MISEP

FIGURE 3.11: Results of image separation

Regularization was achieved by three means:

- The $\mathcal{F}$ network was initialized to perform an identity mapping.
- The $\mathcal{F}$ network was constrained to perform a purely linear separation during the first 100 training epochs. This was achieved by keeping the output weights of the hidden layer units equal to zero during those epochs.

TABLE 3.1: Signal-to-Noise Ratios Achieved by Linear ICA and by MISEP-Based Nonlinear Separation in the Image Separation Example

|  | SOURCE 1 | SOURCE 2 |
| --- | --- | --- |
| Linear separation | 5.2 dB | 10.5 dB |
| Nonlinear separation | 9.3 dB | 13.9 dB |

- The $\mathcal{F}$ network was constrained to be symmetrical (i.e. such that exchanging the inputs would result just in an exchange of the outputs). This constraint was effective because the mixture was known to be symmetrical to a good approximation.

The presentation of this example, here, was necessarily brief. Further details, as well as additional examples, can be found in [10].

## 3.2.2   Nonlinear ICA Through Ensemble Learning

The ensemble learning method of nonlinear ICA that we shall study in this section can be viewed both from a minimum description length (MDL) perspective and from a Bayesian one. There are strong ties between the two perspectives. The two are similar in their mathematical content, but they use interpretations that are quite different from one another. We shall use the MDL perspective here, because it gives more insight into the properties of the ensemble learning method. We shall, however, also make brief references to the Bayesian perspective whenever it's convenient. Most of the published presentations of the method have used the Bayesian approach [61–63, 103, 104]. Two publications that discuss both the Bayesian and the MDL approaches are [48, 102].

Inference based on minimum description length was introduced by Rissanen [85]. Its central idea is that regularities in data can be exploited to compress the data's representation, and that by searching for the model that yields the shortest description of the data one will necessarily find a model that embodies the data's regularities. Tutorials on minimum description length methods can be found in [34, 35]. Here we'll only discuss MDL inasmuch as it applies to the problem at hand.

In nonlinear separation based on ensemble learning, the observed mixtures are represented by means of a nonlinear mixture model of the form

$$x = \mathcal{M}(s, \theta) + n, \tag{3.4}$$

where the mixture vector $x$ results from a nonlinear mixture $\mathcal{M}$ of the sources (the components of $s$), with modeling error $n$ (in the Bayesian framework $n$ is viewed as noise, instead of modeling error). $\theta$ is a vector that gathers all the parameters of the mixture model.

We shall denote by $\bar{x}$ the ordered set of all observation vectors $x_i$ that we have available (with an arbitrarily chosen order, e.g. the order in which they were obtained), and by $\bar{s}$ and $\bar{n}$ the correspondingly ordered sets of source vectors $s_i$ and modeling error vectors $n_i$ respectively. We'll extend the concept of model to these sets of vectors, by writing

$$\bar{x} = \mathcal{M}(\bar{s}, \theta) + \bar{n}. \tag{3.5}$$

Following the MDL approach, we'll assume that we want to describe the mixture observations $\bar{x}$ by encoding the sources $\bar{s}$, the parameters $\theta$ and the modeling error $\bar{n}$, and then using (3.5) to compute the mixture observations from them. The central aspect of MDL is that we seek the representation that corresponds to an encoding with minimal length, measured in terms of the number of coding symbols (e.g. bits). Since what we seek is just the minimal-length representation, the actual encoding doesn't need to be explicitly found. Only its length needs to be computed, so that we can find the $\bar{s}$, $\theta$ and $\bar{n}$ that correspond to the minimal length.

### 3.2.2.1 Coding Lengths and Probability Distributions

Although we don't need to specify the actual encodings of $\bar{s}$, $\theta$ and $\bar{n}$, it is necessary to specify their coding lengths. There are two aspects that we need to examine, regarding these lengths. The first is that the coding of any continuously-valued variable with infinite precision would demand an infinite number of bits, and wouldn't be usable as an optimization criterion. We'll consider encodings with finite resolution, so that the coding lengths become finite. Further on, we'll see that the choice of resolution has virtually no effect on the results of the method.

The second aspect is that it will be convenient to represent coding lengths in an indirect way, by considering associated random variables $\bar{S}$, $\Theta$ and $\bar{N}$, with corresponding probability distributions. It is known that, for a scalar random variable $Z$ with probability density $p(z)$, encoded with resolution $\varepsilon_Z$, the optimal coding length is given by (see [29] Section 9.3, or [35, 48])

$$L(z) = -\log p(z) - \log \varepsilon_Z, \qquad (3.6)$$

where the base of the logarithms defines the unit of measurement of the coding length (e.g. base 2 for bits). This expression is valid if the resolution is fine enough for the density $p(z)$ to be approximately constant within intervals of length $\varepsilon_Z$.

If the random variable is multidimensional, we have

$$L(z) = -\log p(z) - \sum_i \log \varepsilon_{Z_i}, \qquad (3.7)$$

where $\varepsilon_{Z_i}$ is the resolution of the encoding of the $i$-th component of $Z$. Whenever it is necessary to define the coding length for a variable $z$, we will do so by specifying a probability density $p(z)$, which defines the coding length through (3.7).[17] This will allow us to use only valid coding lengths without having to specify the coding scheme [34, 35], and will also allow us to use the tools of probability theory to more easily manipulate coding lengths. In what follows we shall represent $-\sum_i \log \varepsilon_{Z_i}$ by $r_z$, for brevity.

---

[17] In principle we would also need to define the resolutions $\varepsilon_{Z_i}$. However, as we've said before, those resolutions will become relatively immaterial to our discussion.

An aspect that we have glossed over, is that (3.7) may yield a non-integer value, while any actual encoding will involve an integer number of symbols. One way to view Eq. (3.7) is as a continuously valued approximation of that integer length. The continuous approximation has the advantage of being amenable to a more powerful analytical treatment than a discrete one. There are, however, other ways to justify the use of (3.7) (see [29] Section 3.2).

If the data that we are trying to represent come from sampling a random variable, there is an important connection between the true pdf of that variable and the pdf that we use to define the variable's coding length. Let us assume that the random variable $Z$ has pdf $p(z)$, and that we encode it with a length defined by the density $q(z)$. The expected value of the coding length of $Z$ will be

$$E[L_q(Z)] = -\int p(z) \log q(z) dz + r_Z.$$

If we choose $q(z) = p(z)$, the average coding length will be

$$E[L_p(Z)] = -\int p(z) \log p(z) dz + r_Z$$
$$= H(Z) + r_Z.$$

If we use a length defined by a pdf $q(z) \neq p(z)$, the average difference in coding length, relative to $L_p$, is given by

$$E[L_q(Z)] - E[L_p(Z)] = -\int p(z) \log q(z) dz + \int p(z) \log p(z) dz$$
$$= \int p(z) \log \frac{p(z)}{q(z)} dz$$
$$= KLD(p, q).$$

Therefore, the Kullback-Leibler divergence between $p$ and $q$ measures the average excess length resulting from using a coding length defined by $q$ on a random variable whose pdf is $p$. As we know, this KLD is always positive except if $q = p$, in which case it is zero. Consequently, the optimal encoding, in terms of average coding length, is defined by the variable's own pdf.

If $z$ is a vector whose components are i.i.d. samples from a distribution with pdf $p(z)$, the empirical distribution formed by the components of $z$ will be close to $p(z)$, and the coding length of $z$ will be close to minimal if we choose the components' coding lengths according to $p(z)$. Therefore, the choice of the pdf that defines the coding length has a strong relationship with the statistics of the data that are to be encoded.

The representation of coding lengths in terms of probability distributions provides a strong link to the Bayesian approach to ensemble learning. In that approach, the probability distributions of $\bar{S}$, $\Theta$ and $\bar{N}$ correspond to priors over those random variables, and represent our beliefs about those variables' values. In the MDL approach these priors are just convenient means to define coding lengths. However, these priors should be carefully chosen. The reasoning presented in the previous paragraphs shows that the MDL criterion will tend to favor representations which approximately follow the distributions chosen to define the corresponding coding lengths, because those distributions will correspond to the shortest encodings. Those distributions will therefore act as a kind of priors on the representations, even if we are not within a Bayesian framework (and we shall sometimes refer to those distributions as priors, for convenience).

An important aspect of this prior-like effect, in the ICA context, is the following: If two parameters, $z$ and $w$, are encoded with a coding length defined by the product density $p(z, w) = p(z)p(w)$, their coding length will be

$$
\begin{aligned}
L(z, w) &= -\log p(z, w) + r_z + r_w \\
&= -\log p(z) - \log p(w) + r_z + r_w \\
&= L(z) + L(w).
\end{aligned}
$$

Since the coding length of the pair $(z, w)$ is the sum of the coding lengths of the two parameters, this corresponds to coding the two separately from each other. If $z$ and $w$ were obtained from random variables $Z$ and $W$ that are not independent from each other, some coding efficiency will be lost, because some common information will be coded in both variables simultaneously. Consequently, the MDL criterion will tend to favor representations in which the two parameters are statistically independent. Therefore, although the equality $p(z, w) = p(z)p(w)$, above, was not a statement about the independence of the two random variables, but rather a statement about the form in which the variables are coded, it does favor solutions in which the variables are independent.

### 3.2.2.2 Principles of the MDL Approach to Ensemble Learning-Based Separation

Let us collect all the model's arguments from (3.5), $\bar{s}$ and $\boldsymbol{\theta}$, into a single vector $\boldsymbol{\xi}$. We'll also designate this vector as parameter vector, since in MDL the elements of $\bar{s}$ and of $\boldsymbol{\theta}$ are all treated on an equal footing, and thus can all be considered simply as model parameters.

Imagine that we wish to encode a given set of mixture observations $\bar{x}$ and that we choose a certain parameter vector $\boldsymbol{\xi}$ to represent them. This will define a modeling error $\bar{n}$, through (3.5). The coding length of $\bar{x}$ will be

$$
L(\bar{x}) = L(\boldsymbol{\xi}) + L(\bar{n}). \tag{3.8}
$$

$L(\boldsymbol{\xi})$ is defined by the prior $p(\boldsymbol{\xi})$, and $L(\bar{n})$ by the prior $p(\bar{n})$. Equation (3.8) can be written

$$
L(\bar{x}) = -\log p(\boldsymbol{\xi}) - \log p(\bar{n}) + r_{\boldsymbol{\xi}} + r_{\bar{n}}. \tag{3.9}
$$

Once $\boldsymbol{\xi}$ is coded, what remains, to code $\bar{\boldsymbol{x}}$, is coding $\bar{\boldsymbol{n}}$. Therefore, we can write $L(\bar{\boldsymbol{x}}|\boldsymbol{\xi}) = L(\bar{\boldsymbol{n}})$. Furthermore, from (3.5), we see that the resolutions at which the components of $\bar{\boldsymbol{x}}$ are represented are the same as those at which the components of $\bar{\boldsymbol{n}}$ are represented, and consequently $r_{\bar{x}} = r_{\bar{n}}$. Therefore, it is natural to define

$$p(\bar{\boldsymbol{x}}|\boldsymbol{\xi}) = p(\bar{\boldsymbol{n}}), \qquad (3.10)$$

so that

$$
\begin{aligned}
L(\bar{\boldsymbol{x}}|\boldsymbol{\xi}) &= L(\bar{\boldsymbol{n}}) \\
&= -\log p(\bar{\boldsymbol{n}}) + r_{\bar{n}} \\
&= -\log p(\bar{\boldsymbol{x}}|\boldsymbol{\xi}) + r_{\bar{x}}.
\end{aligned}
$$

Equation (3.9) can be written

$$L(\bar{\boldsymbol{x}}) = -\log p(\boldsymbol{\xi}) - \log p(\bar{\boldsymbol{x}}|\boldsymbol{\xi}) + r_{\boldsymbol{\xi}} + r_{\bar{n}}. \qquad (3.11)$$

We wish to choose the representation of $\bar{\boldsymbol{x}}$ with minimum length $L(\bar{\boldsymbol{x}})$. The most obvious solution would correspond to finding the minimum of (3.11), subject to the condition that the modeling equation (3.5) is satisfied for the given $\bar{\boldsymbol{x}}$. There is, however, a form of coding that generally yields a shorter coding length. It is based on the so-called *bits-back* coding method [48, 110], which we shall briefly examine. It uses the fact that (3.5) allows for using any value of $\boldsymbol{\xi}$ for coding any given set of observations $\bar{\boldsymbol{x}}$ (of course, bad choices of $\boldsymbol{\xi}$ will result in large modeling errors).

For a simple example of bits-back coding, assume that we are using a redundant code, in which a certain $\bar{\boldsymbol{x}}$ is represented by two different bit strings, both with the same length of $l$ bits. Of course, choosing the shortest code will yield a coding length of $l$ bits. However, a cleverer scheme is to choose one of the two available code strings according to some other binary information that we need to transmit. In this way we can transmit $\bar{\boldsymbol{x}}$, plus one extra bit of information, in the $l$ bits. This means that, for $\bar{\boldsymbol{x}}$, we will effectively be using $l - 1$ bits, because we will get back the extra bit at the receiver (hence the name of *bits-back* coding). In practice, of course, we don't need to send any extra information, or even to encode the data. We only need to know the coding length, and this reasoning shows that the effective coding length of $\bar{\boldsymbol{x}}$ would be $l - 1$ bits.

In our case we can use any value of $\boldsymbol{\xi}$ for coding any given $\bar{\boldsymbol{x}}$, possibly at the cost of getting large modeling errors. We'll assume that the actual $\boldsymbol{\xi}$ to be used will be chosen at random, with a pdf $q(\boldsymbol{\xi})$, and that its components will be represented with resolutions corresponding to $r_{\boldsymbol{\xi}}$. From (3.11), the average coding length will be

$$\mathrm{E}[L(\bar{\boldsymbol{x}})] = \int q(\boldsymbol{\xi})[-\log p(\boldsymbol{\xi}) - \log p(\bar{\boldsymbol{x}}|\boldsymbol{\xi})]\mathrm{d}\boldsymbol{\xi} + r_{\boldsymbol{\xi}} + r_{\bar{n}}. \qquad (3.12)$$

The amount of information that we can transmit through the choice of $\boldsymbol{\xi}$ is given by the entropy of the discrete random variable obtained from $\boldsymbol{\xi}$, with density $q(\boldsymbol{\xi})$, by discretizing its components with resolutions corresponding to $r_{\boldsymbol{\xi}}$. For fine enough resolutions, this entropy is given by (see [29] Section 9.3)

$$H_q(\boldsymbol{\xi}) = -\int q(\boldsymbol{\xi}) \log q(\boldsymbol{\xi}) d\boldsymbol{\xi} + r_{\boldsymbol{\xi}}.$$

Therefore the effective bits-back coding length of $\bar{x}$, which we represent by $\hat{L}_q(\bar{x})$, is given by

$$\hat{L}_q(\bar{x}) = \mathrm{E}[L(\bar{x})] - H_q(\boldsymbol{\xi})$$
$$= \int q(\boldsymbol{\xi})[-\log p(\boldsymbol{\xi}) - \log p(\bar{x}|\boldsymbol{\xi}) + \log q(\boldsymbol{\xi})] d\boldsymbol{\xi} + r_{\bar{n}}.$$

Note that, in this equation, the term $r_{\boldsymbol{\xi}}$, relative to the resolution of the representation of $\boldsymbol{\xi}$, has been canceled out. This means that bits-back coding allows us to encode the model parameters with as fine a resolution as we wish, without incurring any penalty in terms of coding length.

The latter equation can be transformed as

$$\hat{L}_q(\bar{x}) = \int q(\boldsymbol{\xi}) \log \frac{q(\boldsymbol{\xi})}{p(\boldsymbol{\xi})p(\bar{x}|\boldsymbol{\xi})} d\boldsymbol{\xi} + r_{\bar{n}} \tag{3.13}$$

$$= \int q(\boldsymbol{\xi}) \log \frac{q(\boldsymbol{\xi})}{p(\boldsymbol{\xi}|\bar{x})p(\bar{x})} d\boldsymbol{\xi} + r_{\bar{n}} \tag{3.14}$$

$$= KLD[q(\boldsymbol{\xi}), p(\boldsymbol{\xi}|\bar{x})] - \log p(\bar{x}) + r_{\bar{n}}, \tag{3.15}$$

where $p(\bar{x})$ and $p(\boldsymbol{\xi}|\bar{x})$ are defined, consistently with probability theory, as

$$p(\bar{x}) = \int p(\bar{x}|\boldsymbol{\xi})p(\boldsymbol{\xi}) d\boldsymbol{\xi}$$
$$p(\boldsymbol{\xi}|\bar{x}) = \frac{p(\bar{x}|\boldsymbol{\xi})p(\boldsymbol{\xi})}{p(\bar{x})}.$$

In Eq. (3.15), the term $-\log p(\bar{x})$ doesn't depend on $q(\boldsymbol{\xi})$. On the other hand, the KLD in that equation has a minimum (equal to zero) for $q(\boldsymbol{\xi}) = p(\boldsymbol{\xi}|\bar{x})$. This shows that the optimal choice for $q(\boldsymbol{\xi})$ would be $p(\boldsymbol{\xi}|\bar{x})$. However, the latter probability is often hard to compute in real nonlinear separation problems. In the ensemble learning method we use a distribution $q(\boldsymbol{\xi})$ which is an approximation to $p(\boldsymbol{\xi}|\bar{x})$. Equation (3.15) shows that, by using that approximation, we incur a penalty equal to $KLD[q(\boldsymbol{\xi}), p(\boldsymbol{\xi}|\bar{x})]$, and that $q(\boldsymbol{\xi})$ should be selected so as to minimize that KLD within the chosen approximation family. Note, however, that the actual optimization is performed using (3.13), and not (3.14). This is because all the terms in the former equation can be computed, while it would be hard to obtain $p(\boldsymbol{\xi}|\bar{x})$.

### 3.2.2.3 Practical Aspects

We are now in a position to give an overview of the ensemble learning method of nonlinear source separation. The overview will cover the most important aspects, but several details will have to be skimmed over. The reader is referred to [102, 104] for more complete presentations.

The method involves a number of choices and approximations, most of which are intended at making it computationally tractable:

- The mixture model $\mathcal{M}$ in (3.4) is chosen to be a multilayer perceptron with a single hidden layer of sigmoidal units and with an output layer formed by linear units. The weights and biases of this network are components of the parameter vector $\boldsymbol{\theta}$.

- The unknowns are encoded independently from one another, and therefore their prior is highly factored,[18]

$$p(\bar{s}, \boldsymbol{\theta}, \boldsymbol{n}) = \prod_{i,j} p(s_{ij}) \prod_k p(\theta_k) \prod_{l,m} p(n_{lm}), \qquad (3.16)$$

where $s_{ij}$ is the $j$-th sample of the $i$-th source, and a similar convention applies to $n_{lm}$.

- The priors of the parameters and of the noise samples, $p(\theta_k)$ and $p(n_{lm})$ respectively, are taken to be Gaussian. The means and variances of these Gaussians are not fixed a priori and, instead, are also encoded. For this reason these means and variances are called *hyperparameters*. They are further discussed ahead. From (3.10), $p(\bar{x}|\boldsymbol{\xi})$ factors as a product of Gaussians, because $p(\bar{n})$ does.

- The priors $p(s_{ij})$ can be simply chosen as Gaussians. In this case, only linear combinations of the sources are normally found. This method is called nonlinear factor analysis (NFA), and is often followed by an ordinary linear ICA operation, to extract the independent sources from these linear combinations.

  The source priors can also be chosen as mixtures of Gaussians, in order to have a good flexibility in modeling the source distributions, in an attempt to directly perform nonlinear ICA. However, this increased flexibility often does not lead to a full separation of the sources, as explained ahead. For that reason, it is more common to use the approach of performing NFA (using simple Gaussian priors) followed by linear ICA, rather than trying to directly perform nonlinear ICA, using mixtures of Gaussians for the priors.

  The parameters of the Gaussians (means and variances) or of the mixtures (weights, means and variances of the mixture components) are hyperparameters.

---

[18]In the second example presented ahead, with a dynamical model of the sources, the factorization of $p(\bar{s})$ doesn't apply, being replaced by the dynamical model.

- The hyperparameters, which are also members of the parameter vector $\boldsymbol{\theta}$, are chosen to be the same for large numbers of parameter distributions, in order to reduce the number of hyperparameters. For example, the hyperparameters for all the Gaussians, or for all the components of the mixtures of Gaussians that represent the sources are the same. More specifically, all component means have the same prior Gaussian distribution, and the same happens with all component variances (and with all component weights, if mixtures are used). Therefore there are only four or six hyperparameters for all the prior source distributions (two for the means, two for the variances and possibly two for the weights). Furthermore, the weights of each layer of the MLP implementing $\mathcal{M}$ all have the same prior distribution, which is Gaussian. Therefore there are only two hyperparameters for all the weights of each layer. In this way the hyperparameters are reduced to a small number.

- The hyperparameters are given very broad prior distributions (based on Gaussians with very large variances) in order to make them very unrestrictive. If we had prior information about the sources or the mixture process, it would be useful to incorporate that information into the prior distributions. But usually we don't have such information, and then it is preferable to make those distributions as unrestrictive as possible.

- The approximator $q(\boldsymbol{\xi})$ is factored as

$$q(\boldsymbol{\xi}) = \prod_{i,j} q(s_{ij}) \prod_k q(\theta_k), \tag{3.17}$$

where the $q(\cdot)$ on both factors of the right hand side are Gaussians.[19] Therefore one only needs to estimate their means and variances.

Since $q(\boldsymbol{\xi}) \approx p(\boldsymbol{\xi}|\bar{\boldsymbol{x}})$, the factorization in (3.17) will tend to make the sources mutually independent, given $\bar{\boldsymbol{x}}$. This bias for independence is necessary to make the model tractable.[20]

- Those means and variances are estimated by minimizing the coding length $\hat{L}_q(\bar{\boldsymbol{x}})$, given by (3.13). The resolutions that are chosen for representing the modeling error's components affect this length only through the additive term $r_{\bar{n}}$, which doesn't affect the position of the minimum. That term can, therefore, be dropped from the minimization.

---

[19]The densities $q(s_{ij})$ are taken as mixtures of Gaussians if one is doing nonlinear ICA, and not just NFA.

[20]This independence bias has some influence on the final result of separation, sometimes yielding a linear combination of the sources instead of completely separated ones, even if one uses mixtures of Gaussians for the source priors [57]. This is why some authors prefer to use ensemble learning to perform only nonlinear factor analysis (NFA), extracting only linear combinations of the sources, and then to use a standard linear ICA method to separate the sources from those linear combinations. This is what is done in the first application example, presented ahead.

Consequently, the objective function that is actually used is

$$C = \int q(\xi) \log \frac{q(\xi)}{p(\xi)p(\bar{x}|\xi)} d\xi. \qquad (3.18)$$

As a side note, and since the KLD in (3.15) is non-negative, we conclude from (3.13) and (3.15) that the objective function $C$ gives an upper bound for $-\log p(\bar{x})$. In the Bayesian framework, $p(\bar{x})$ is the probability density that the observed data would be generated by the model (3.5), for the specific form of $\mathcal{M}$ and for the specific priors that were chosen. This probability is often called the *model evidence*, and the value of the objective function can be used to compute a lower bound for it.

- The approximator $q(\xi)$ and the prior $p(\xi)$ factor into products of large numbers of simple terms – see (3.17) and (3.16). The posterior $p(\bar{x}|\xi)$ is approximated by a product of Gaussians,

$$p(\bar{x}|\xi) \approx \prod_{ij} p(x_{ij}). \qquad (3.19)$$

The factorization and the presence of the logarithms in (3.18) leads the objective function to become a sum of a large number of relatively simple terms which can all be computed, either exactly or approximately.

These choices and approximations make it possible to compute closed form expressions of the partial derivatives of the objective function relative to the parameters (means and standard deviations) of the approximator $q(\xi)$. The partial derivatives are set equal to zero, and this leads to equations that are simple enough to be solved directly (for some of the equations, only approximate solutions can be found). This allows the estimation of the parameters, for one subset of the $q(\xi_i)$ at a time, taking the parameters for the other $\xi_i$ as constant. The procedure iterates through these estimations until convergence. This procedure, although computationally heavy, is nevertheless faster than gradient descent on the KLD. Conjugate gradient optimization [39] also gives good results.[21]

Once the minimum is found, the density

$$q(\bar{s}) \approx p(\bar{s}|\bar{x}) \qquad (3.20)$$

yields an estimate of the posterior distribution of the sources. Due to the factorization of $q(\xi)$, one can directly obtain the approximate posterior of the sources, $q(\bar{s})$, from the complete posterior $q(\xi)$, by keeping only the terms relative to $\bar{s}$.

---

[21]H. Valpola, private communication, 2005.

Within an MDL framework, the maximum of $q(\bar{s})$ yields the estimated sources. In fact, from (3.11),

$$L(\bar{x}) = -\log p(\xi|\bar{x}) - \log p(\bar{x}) + r_\xi + r_{\bar{n}}.$$

Since $p(\bar{x})$, $r_\xi$ and $r_{\bar{n}}$ don't depend on $\xi$, they can be dropped from the optimization. Furthermore,

$$p(\xi|\bar{x}) \approx q(\xi) \tag{3.21}$$
$$= q(\bar{s})q(\boldsymbol{\theta}). \tag{3.22}$$

Therefore, the minimum length encoding will correspond (within these approximations) to the $\bar{s}$ and $\boldsymbol{\theta}$ that maximize the corresponding terms in (3.22).[22] If $q(\bar{s})$ is represented as a product of Gaussians on the various $s_{ij}$, as is normally done, the MDL estimate of $\bar{s}$ will simply correspond to the estimated means of those Gaussians.

Within a Bayesian framework, $p(\bar{s}|\bar{x})$ is interpreted as the true posterior of the sources, in the statistical sense, and $q(\bar{s})$ is an approximation to it. Therefore it can be used, for example, for computing means or MAP estimates, or for taking decisions that depend on $\bar{s}$, as is normally done with statistical distributions.

### 3.2.2.4 Model Selection

The optimization method that we have presented is based on a single model structure, because the architecture of the model's MLP is assumed to be fixed. However, the minimum description length principle yields a simple and natural criterion for comparing different model structures: the model that achieves the lowest description length is best. But without some guiding lines, the search for the best model structure would be blind.

Interestingly, the MDL framework also provides information that can guide the search for better model structures. After optimizing the parameters of a model based on an MLP with a certain architecture, we can find the coding length of each weight. Weights that are coded with very few bits probably are rather unimportant to the network's operation, and are candidates for pruning. The pruned network can be optimized again, and the resulting coding length compared with the one of the original network, to decide which one is best. The pruning procedure can be iterated, if desired.

An important property of ensemble learning is that it inherently avoids overfitting. The MDL criterion is a form of Occam's razor [34, 74]. It automatically seeks the simplest explanation for the data to be modeled, and therefore avoids finding too complex a model when the available data do not warrant it. An interesting example, in a linear separation context, is given

---

[22]Incidentally, these are also the maximum a posteriori (MAP) estimates of $\bar{s}$ and $\boldsymbol{\theta}$, within the approximation (3.21).

in [48]. In that example, a relatively small amount of training data led FastICA to find spurious spiky sources, due to overfitting, while ensemble learning found relatively good approximations of the actual sources.

### 3.2.2.5  Extensions of the Method

The basic ensemble learning method of nonlinear ICA has been extended in two main directions. One involves the use of dynamical models of the temporal structure of the source signals [106], and will be examined in some more detail ahead, in the context of an example. The other direction involves the use of standard blocks that allow the easy construction of hierarchical models, where some variables can control the variances of other variables, for example, or can control the switching among several random variables. These blocks also allow the construction of MLP-like structures, although the kinds of nonlinear activation functions that can be used are limited [107, 108]. The advantage is that these models involve fewer approximations, and are often faster to compute than the basic MLP-based model described above. These models have been applied to the processing of magneto-encephalographic signals in [105].

In [46, 50] a faster optimization method has been proposed, which can be used both for ensemble learning and for certain other classes of problems. In [49] an improvement of the basic nonlinear separation method was proposed, which avoids some of the approximations of the basic method, and is both more reliable and more accurate. Another interesting result was given in [47], where it was shown that kernel PCA [90] provides an effective means to initialize the optimization of nonlinear separation systems based on ensemble learning. Ensemble learning has also been used to separate post-nonlinear mixtures [57]. In that context it showed to be able to perform the separation even when not all individual nonlinearities are invertible, in situations with more mixture components than sources, as long as the nonlinear mapping, as a whole, is still invertible.

### 3.2.2.6  Application Examples

We shall present two examples of the application of ensemble learning, the first one corresponding to an instantaneous model without any dynamical part, and the second one involving a dynamical model. Another interesting example can be found in [62]. In that example, 10 features were nonlinearly extracted from a 30-component time series of real-life data from a pulp mill. Those features were shown to allow the reconstruction of the original data with good accuracy.

**Non-Dynamical Model**   In this case the sources were eight random signals, four of which had supergaussian distributions, the other four being subgaussian. These signals were nonlinearly mixed by a multilayer perceptron with one hidden layer, with $\sinh^{-1}$ activation functions. This

**FIGURE 3.12:** Results of the first example of separation using ensemble learning. In each scatter plot, the horizontal axis corresponds to a true source, and the vertical axis to an extracted source. The sources in the top row were supergaussian, and those in the lower row were subgaussian.

mixing MLP had 30 hidden units and 20 output units, and its weights were randomly chosen. Note that, since 20 mixtures were observed and there were only 8 sources, this was not a square mixing situation, unlike other ones that we have seen in this book. Gaussian noise, with an SNR of 20 dB, was added to the mixture components.

The analysis was made using nonlinear factor analysis (i.e. Gaussian priors were used for the sources), using eight sources and a mixture model consisting of an MLP with a single hidden layer of 50 units, with tanh activation functions. Since the source priors were Gaussian, the sources themselves were not extracted in this step, but only linear combinations of them. In a succeeding step, the FastICA method of linear ICA (see Section 2.4.2) was used to perform separation from these linear combinations. Figure 3.12 shows the scatter plots of the extracted components (after the FastICA step) versus the corresponding sources, and confirms that the sources were extracted to a very good accuracy. This is also shown by the average SNR of the extracted sources relative to the true ones, which was 19.6 dB. The extraction was slow, though, having taken 100,000 training epochs.

**Separation with a Dynamical Model**    Our second example corresponds to the dynamical extension of ensemble learning [106]. In this example, the mixture observations are still modeled through (3.5), but the sources are not directly coded. Instead, they are assumed to be observations that are sequential in time, and are coded through a dynamical model

$$s(t) = \mathcal{G}[s(t-1), \phi] + m(t), \qquad (3.23)$$

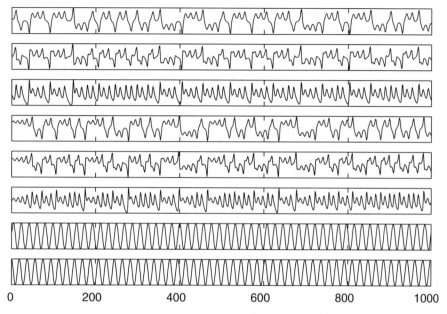

**FIGURE 3.13:** Separation by ensemble learning with a dynamical model: The eight sources.

where $\mathcal{G}$ is a nonlinear function parameterized through $\phi$.[23] Therefore, $s(t)$ is coded through $\phi$, $m(t)$ and $s(1)$. The $m(t)$ process is often called the *innovation process*, since it is what drives the dynamical model. The variables to be encoded are the parameters of the nonlinear mappings ($\theta$ and $\phi$), the innovation process $m(t)$, the initial value $s(1)$ and the modeling error $n(t)$. All of these variables were separated into their scalar components, and each component had a Gaussian prior. These Gaussians involved some hyperparameters that were given very broad distributions, as usual.

For generating the test data for this example, the source dynamical processes consisted of two independent chaotic Lorenz processes [70] (with two different sets of parameters) and a harmonic oscillator. A Lorenz process has three state variables and an oscillator has two, meaning that the system had a total of eight state variables, forming three independent dynamical processes. Figure 3.13 shows a plot of the eight source state variables.

For producing the mixture observations, these eight state variables were first linearly projected into a five-dimensional space. Due to this projection into a lower-dimensional space, the behavior of the system could not be reconstructed without learning its dynamics, because five variables do not suffice to represent the eight-dimensional state space. These five projections were then nonlinearly mixed, by means of an MLP with a single hidden layer with $\sinh^{-1}$

---

[23]We have slightly changed the notation, in this equation, using $s(t)$ instead of $\bar{s}$, and $m(t)$ instead of $\bar{m}$, to emphasize the temporal aspect of the dynamical model.

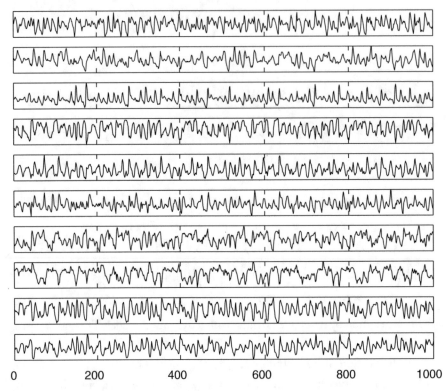

FIGURE 3.14: Separation by ensemble learning with a dynamical model: The ten nonlinear mixture components.

nonlinearities and randomly chosen weights, producing a 10-dimensional observation vector $x(t)$, shown in Fig. 3.14.

The analysis was performed using a model corresponding to Eqs. (3.5) and (3.23), where $\mathcal{G}$ and $\mathcal{M}$ were modeled by multilayer perceptrons with 30 hidden units each, with tanh nonlinearities and with linear output units. The best results were obtained with a model with nine sources. However, after training, one of the sources was essentially unused (it was almost constant). The training procedure involved two successive phases. In the first phase, lasting for the first 500 epochs, a 50-component vector consisting of $x(t) \cdots x(t-4)$ was used as the mixture observation, to provide a suitable embedding, as is often done with dynamical systems [40]. This allowed the system to learn an initial, still rather crude dynamical model. In the second phase, after these initial epochs, only the vector $x(t)$ was used as mixture observation.

Learning of a good model took a total of 500,000 epochs. This long time should be no surprise, since learning a good model of a chaotic process often is a very hard task. Figure 3.15 illustrates the quality of the results. In that figure, the first 1000 samples correspond to sources that were estimated from observed mixtures $x(t)$. The last 1000 samples correspond to a "continuation" of the process, using only the dynamical model $s(t) = \mathcal{G}[(s(t-1), \phi]$, with no innovation. The fact that the correct dynamical behavior of the sources has been obtained in this continuation shows that a very good dynamical model was learned. In fact, in chaotic processes, even small deviations from the correct model often lead to a rather different attractor, and therefore to a rather different continuation.

For comparing with more classical methods, several attempts were made to learn a dynamical model operating directly on the observations, and using MLPs of several different configurations for implementing its nonlinear dynamics. None of these attempts was able to yield the correct behavior, in a continuation of the dynamical process.

This example is necessarily rather complex, and here we could only give an overview. One aspect that we have glossed over, is that only a nonlinearly transformed state space was learned by the method. This space was then mapped, by means of an MLP, to the original state space, in order for the recovery of the correct dynamics to be checked. What is shown in Fig. 3.15 is the result of that mapping of the learned dynamics into the original state space. However, even in the state space learned by the method, the three dynamical processes were separated. And the fact that a transformed state space was learned has little relevance, since there is not a particular space that can claim being the "main" state space, in a nonlinear dynamical process. For further details see [106].

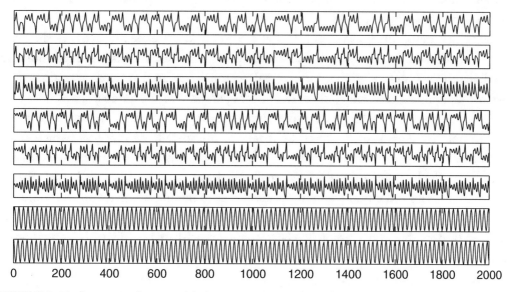

FIGURE 3.15: Separation by ensemble learning with a dynamical model: Extracted sources. The first 1000 samples correspond to sources estimated using mixture observations. The last 1000 samples correspond to a continuation, by iteration of the dynamical model $s(t) = g[(s(t-1)]$.

In an interesting development, it was shown that this method can also accurately detect changes in the properties of dynamical processes [57, 58].

### 3.2.3   Kernel-Based Nonlinear Separation: kTDSEP

The nonlinear separation method that we shall now discuss was proposed by Harmeling *et al.* [37, 38]. It is based on a rather simple idea: If we make a nonlinear mapping from the space of the mixtures (that we shall designate by $\mathcal{X}$) into some other space $\hat{\mathcal{X}}$ (usually called *feature space*), and then perform a linear separation in that space, that will correspond to a nonlinear separation in the original space. Similar methods have been used, with very good results, in other classes of problems. For example, the very successful *support vector machines* [24, 76, 89] are nonlinear classifiers or nonlinear regressors that first make a nonlinear mapping from the original pattern space into a feature space, and then perform linear classification or linear regression in that space. *Kernel PCA* [90] is a form of nonlinear PCA in which the data are first nonlinearly mapped into a feature space, and linear PCA is then performed in that space. A good review of kernel-based methods is given in [76].

Although the basic idea behind kernel-based nonlinear separation is simple, applying it in a useful way involves rather careful, and sometimes nonobvious, choices. One basic issue is how to choose the nonlinear mapping to be used. This will determine, in an indirect way, what class of nonlinear transformations can be performed on the original data. That class should include the transformation that we wish to perform (or at least a good approximation to it). However, in many kinds of problems, including nonlinear source separation, we do not explicitly know, *a priori*, which transformation we wish to apply: we only know some properties of the desired result. The solution that is normally adopted is to use a nonlinear transformation into a very high dimensional (or even infinite-dimensional) feature space, so that linear operations in that space correspond to a very wide class of nonlinear operations in the original space.

Working in the very high or infinite dimensional space $\hat{\mathcal{X}}$ raises the issue of computational complexity. This is often solved by means of the so-called *kernel trick*, which enables us to indirectly perform certain operations in the high-dimensional $\hat{\mathcal{X}}$ by actually performing operations in the low-dimensional $\mathcal{X}$. We shall start by briefly discussing the kernel trick, because this makes a good introduction to the kernel-based nonlinear separation method. Before proceeding, however, we should note that there is an ICA method called *kernel ICA* [18] which is a *linear* ICA method, only remotely related to the kernel-based nonlinear ICA method that we are discussing in this section. Both methods employ the kernel trick, but for rather different purposes. However, the similarity of the names may cause some confusion.

Let us denote by $x_i$ generic vectors of the low-dimensional mixture space $\mathcal{X}$, and by $\hat{x}_i$ the corresponding mapped vectors in the high-dimensional feature space $\hat{\mathcal{X}}$. Assume that the operations that we need to perform in $\hat{\mathcal{X}}$ can all be expressed exclusively in terms of inner products of vectors of that space. Also assume that, for any two vectors of $\mathcal{X}$, $\boldsymbol{x}_1$ and $\boldsymbol{x}_2$, the inner product of their images in $\hat{\mathcal{X}}$ can be expressed as a relatively simple function of the original vectors $\boldsymbol{x}_1$ and $\boldsymbol{x}_2$:

$$\hat{\boldsymbol{x}}_1 \cdot \hat{\boldsymbol{x}}_2 = k(\boldsymbol{x}_1, \boldsymbol{x}_2). \tag{3.24}$$

The function $k(\cdot, \cdot)$ is then called the kernel of the mapping from $\mathcal{X}$ into $\hat{\mathcal{X}}$. If (3.24) holds, the whole algorithm that we wish to implement can be performed in the low-dimensional space $\mathcal{X}$ by replacing the inner products in $\hat{\mathcal{X}}$ with kernel evaluations in $\mathcal{X}$. This will avoid performing operations in the high-dimensional space $\hat{\mathcal{X}}$.

This may seem too far fetched at first sight. After all, most nonlinear mappings will not have a corresponding kernel that obeys (3.24). However, what is done in practice is to choose only mappings that have a corresponding kernel. In fact, the issue is often reversed: one chooses the kernel, and that implicitly defines the nonlinear mapping that is used. The important point is that it is not hard to find kernels corresponding to mappings that yield very wide classes of nonlinear operations in $\mathcal{X}$, when linear operations are performed in $\hat{\mathcal{X}}$.

The most straightforward application of these ideas to nonlinear ICA would consist of performing linear ICA in the feature space $\hat{\mathcal{X}}$, and then mapping the obtained components back to the original space $\mathcal{X}$. This is not possible, however, because it has not been possible to express linear ICA only in terms of inner products, and these are the only operations that can be efficiently performed in $\hat{\mathcal{X}}$. And even if this could be done, there would probably be the additional problem that working (albeit indirectly) in such a high-dimensional space could easily lead to badly conditioned systems, raising numerical instability issues.

For these reasons, the approach that is taken in kernel-based nonlinear ICA is different: we first linearly project the data from the feature space $\hat{\mathcal{X}}$ into a medium-dimensional *intermediate space* $\tilde{\mathcal{X}}$, and then perform linear ICA in that space. Normally, linear ICA in this lower dimensional space already is numerically tractable. The dimension of $\tilde{\mathcal{X}}$ (call it $d$) is often chosen in a way that depends on the existing data, but for a typical two-source problem it could be around 20.

The projection from the feature space into the intermediate space should be made in such a way that the most important information from the feature space is kept. This projection can be performed in several ways. The one that immediately comes to mind is the use of PCA in $\hat{\mathcal{X}}$. This amounts to using the kernel-PCA method that we mentioned earlier [90]. That method uses only inner products in $\hat{\mathcal{X}}$, and can therefore be efficiently implemented by means of the kernel trick. Other alternatives that also use only inner products include the use of random sampling or of clustering [38].

Once the data have been projected into the intermediate space $\tilde{\mathcal{X}}$, it is possible to perform linear ICA in that space through several standard methods. The method that has been used in practice is TDSEP (Section 2.3), which, we recall, involves the simultaneous decorrelation of the extracted components, with a set of different time delays. Because the kernel-based nonlinear separation method involves the use of kernels and TDSEP, it is often called kTDSEP.

A difficulty that arises, and that might not be obvious at first sight, is that linear ICA, performed in the intermediate space, yields a number of components equal to the dimension of that space, a number which is significantly larger than the number of sources $n$. This might seem contradictory because, strictly speaking, there can be no more independent components than the number of sources, which is $n$. However, TDSEP just performs a joint diagonalization of the various delayed correlation mixtures that are involved. If a perfect diagonalization cannot be achieved, it finds a diagonalization which is as perfect as possible, according to its cost function. This means that it will always find the $d$ components that are, in some sense, "as independent as possible." But most of these components cannot actually be independent from one another. This raises the need to select the $n$ components that are as mutually independent as possible, and as close to the original sources as possible.

The procedure that has been used for performing this selection is heuristic but appears to work well in practice. Briefly speaking, it consists of applying the nonlinear separation method again to the $d$ components resulting from the first separation, and then selecting the components that have changed the least from the first to the second separation. The rationale is that the true sources will tend to exhibit less variability than "spurious" components from the first to the second separation.

### 3.2.3.1 Examples
We shall present two examples of the separation of nonlinear mixtures through kTDSEP.

**First Example**    The source signals were two sinusoids with different frequencies (2000 samples of each). Fig. 3.16(a) shows these signals, and Fig. 3.17(a) shows the corresponding scatter plot.

These signals were nonlinearly mixed according to

$$x_1 = e^{s_1} - e^{s_2}$$
$$x_2 = e^{-s_1} + e^{-s_2}. \tag{3.25}$$

Fig. 3.16(b) shows the mixture signals, and Fig. 3.17(b) shows the corresponding scatter plot. We can see that the mixture performed by (3.25) was significantly nonlinear.

Figs. 3.16(c) and 3.17(c) show the result of linear ICA. Both figures show that it was not able to perform a good separation, as expected. Fig. 3.17(c) shows that linear ICA just rotated the joint distribution. Naturally, it could not undo the nonlinearities introduced by the mixture.

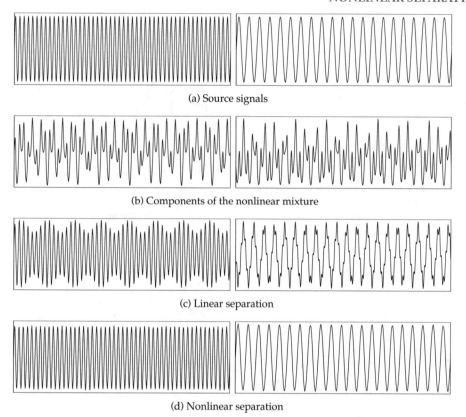

(a) Source signals

(b) Components of the nonlinear mixture

(c) Linear separation

(d) Nonlinear separation

**FIGURE 3.16:** Signals corresponding to the first example of separation through kTDSEP

Nonlinear separation through kTDSEP was performed using a so-called polynomial kernel, of the form

$$k(\boldsymbol{a}_1, \boldsymbol{a}_2) = (\boldsymbol{a}_1^{\mathrm{T}}\boldsymbol{a}_2 + 1)^9.$$

This kernel generates a feature space $\hat{\mathcal{X}}$ spanned by all the monomials of the form $(x_1)^m(x_2)^n$, with $m + n \leq 9$, where $x_1$ and $x_2$ are the two mixture components. This space has a total of 54 dimensions. This space was then reduced, through a clustering technique, to 20 dimensions, thus forming the intermediate space $\tilde{\mathcal{X}}$. Linear ICA was performed on this intermediate space by the TDSEP technique, yielding 20 components. From these, the ones corresponding to the sources were extracted, as indicated above, by rerunning the algorithm on the set of 20 components and selecting the two that were least modified by this second separation. The results are shown in Figs. 3.16(d) and 3.17(d). We can see that a virtually perfect separation was achieved.

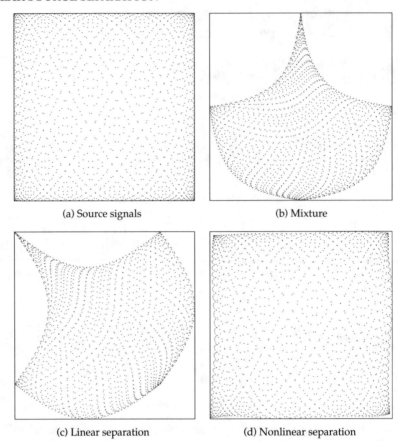

(a) Source signals          (b) Mixture

(c) Linear separation        (d) Nonlinear separation

**FIGURE 3.17:** Scatter plots from the first example of separation through kTDSEP

**Second Example** The sources were two speech signals, each with a length of 20 000 samples. Fig. 3.18(a) shows these signals, and Fig. 3.19(a) shows their joint distribution. The latter figure shows that the sources were strongly supergaussian, as normally happens with speech signals.

For performing the nonlinear mixture, the signals were first both scaled to the interval $[-1, 1]$ and were then mixed according to

$$x_1 = -(s_2 + 1) \cos(\pi s_1)$$
$$x_2 = (s_2 + 1) \sin(\pi s_1).$$

We can describe this mixture in the following way: The mixture space is generated in polar coordinates. Source $s_1$ controls the angle, while $(s_2 + 1)$ controls the distance from the center. Figs. 3.18(b) and 3.19(b) show the mixture results.[24]

---

[24]Due to the shape of the mixture scatter plot, this has become commonly known as the "euro mixture."

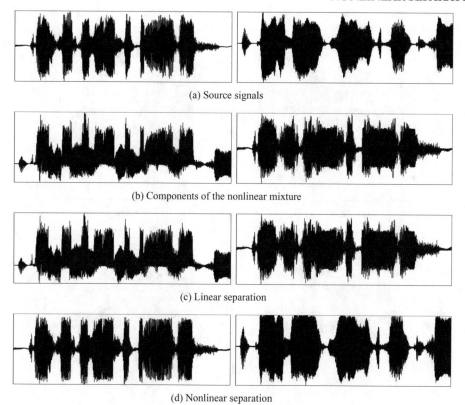

(a) Source signals

(b) Components of the nonlinear mixture

(c) Linear separation

(d) Nonlinear separation

**FIGURE 3.18:** Signals corresponding to the second example of separation through kTDSEP

Linear ICA naturally was unable to perform an adequate separation, as shown in Figs. 3.18(c) and 3.19(c). As in the previous example, it just performed a rotation of the mixture distribution.

Nonlinear separation was performed using a Gaussian kernel of the form

$$k(\boldsymbol{a}_1, \boldsymbol{a}_2) = e^{-||\boldsymbol{a}_1 - \boldsymbol{a}_2||^2}.$$

This kernel induces an infinite-dimensional feature space $\hat{\mathcal{X}}$. The dimensionality reduction to an intermediate space $\tilde{\mathcal{X}}$ (again of dimension 20) was performed, in this case, by means of a random sampling technique. Linear ICA, through the TDSEP method, was then performed on this intermediate space. The two components corresponding to the extracted sources were again selected by rerunning the algorithm on these separation results and selecting the components that changed the least from the first to the second separation.

Figs. 3.18(d) and 3.19(d) show the separation results. Once again, we see that a much better source recovery was achieved than with linear ICA. This fact was confirmed by computing

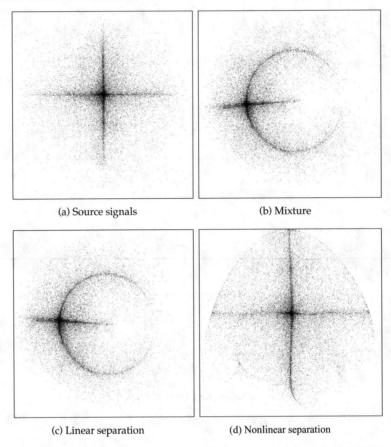

(a) Source signals    (b) Mixture

(c) Linear separation   (d) Nonlinear separation

**FIGURE 3.19:** Scatter plots from the second example of separation through kTDSEP

the signal-to-noise ratios of the results of both linear and nonlinear separation, relative to the original sources. Table 3.2 shows the results and confirms the very large improvement obtained with kTDSEP, relative to linear ICA.

We have shown examples of the separation of nonlinear mixtures of just two sources. However, kTDSEP has been shown to be able to efficiently separate mixtures of up to seven sources [38].

| **TABLE 3.2:** Signal-to-Noise Ratios of the Results of Linear ICA and kTDSEP | | |
|---|---|---|
| | **SOURCE 1** | **SOURCE 2** |
| Linear ICA | 5.4 dB | −7.0 dB |
| kTDSEP | 13.2 dB | 18.1 dB |

### 3.2.4   Other Methods

In this section we shall give an overview of other methods that have been proposed for performing nonlinear source separation, and we shall simultaneously try to give an historical perspective of the field. The number of different methods that we shall mention is large, and therefore we can only make a brief reference to each of them.

A very early result on nonlinear source separation was published by Darmois in 1953 [30]. He showed the essential ill-posedness of unconstrained nonlinear ICA; i.e., that it has an infinite number of solutions that are not related to one another in a simple way.

One of the first nonlinear ICA methods was proposed by Schmidhuber in 1992 [88]. It was based on the idea of extracting, from the observed mixture, a set of components such that each component would be as unpredictable as possible from the set of all other components. This justified the method's name, *predictability minimization*. The components were extracted by an MLP, and the attempted prediction was also performed by MLPs. While the basic idea was sound, the method was computationally heavy, and hard to apply in practice.

In the same year, another nonlinear ICA method was proposed by Burel [23]. It was based on the minimization of a cost function that was a smoothed version of the quadratic error between the true distribution $p(\boldsymbol{y})$ and the product of the marginals, $\prod_i p(y_i)$. The method used an MLP as a separator. The cost function was expressed as a series involving the moments of the extracted components. The series was then truncated and the moments were estimated on the training set. Backpropagation was used to compute the gradient of the cost function for minimization. The method was demonstrated to work on an artificially generated nonlinear mixture, both without and with noise. While no explicit regularization conditions were mentioned, the smoothing of the error, the series truncation, and the fact that a very small MLP was used (with one hidden layer with just two units) were implicit regularizing constraints that allowed the method to cope with the indetermination of nonlinear ICA for the mixture that was considered in the examples.

In 1995, Deco and Brauer proposed a method based on the minimization of the mutual information of the extracted components [31]. The method was restricted to volume-conserving nonlinear mixtures (also called information-preserving mixtures), and the separation was performed by an MLP with a special structure. The probability densities needed for the estimation of the mutual information were approximated by truncated series based on cumulants. The indetermination inherent to nonlinear ICA was handled by the restriction to volume-conserving mixtures and also by the inherent regularization corresponding to the truncation of the cumulant-based series. Variants of the method were proposed in [81, 82].

Also in 1995, Hecht–Nielsen proposed a method (replicator neural networks) to find a mapping from observed data to data that are uniformly distributed within a hypercube [41]. Coordinates parallel to the hypercube's edges (which were designated *natural coordinates* of the original data) actually were independent components, although that was not noted at the time. The method was mainly intended for dimensionality reduction (and therefore, in our language, for problems with more mixture components than sources), but there is nothing that prevents it from being applied to square problems.

In 1996, Pajunen [77] proposed a method based on the use of self-organizing maps (SOM) [39]. SOMs learn a mapping of the mixture into a grid of "centers" that are uniformly distributed within a hypercube. The hypercube's coordinates become the extracted components, and are approximately independent from one another, because the distribution is approximately uniform within the hypercube. The SOM learning algorithm provides an inherent regularization that can, to a certain extent, deal with the indetermination of nonlinear ICA. The map that is obtained is discrete, because it is made into the discretely located centers. An adequate interpolation is performed to handle intermediate mixture vectors. Pajunen's work was followed by [68,69], and the method has been applied in [36] to an image denoising problem. An approach based on generative topographic mapping, which is a more principled form of self-organizing mapping, was proposed in [79].

Also in 1996, Marques and L. Almeida [71] proposed a new class of objective functions for achieving statistical independence. These objective functions had the advantage of being able to enforce independence without any approximations and without the need to estimate any probability density functions, all computations being based directly on the training set. The use of these objective functions was illustrated for linear ICA and for very small examples of nonlinear ICA. Extension of their use to nonlinear ICA situations of realistic size has been difficult, however.

In 1997, Fisher and Principe [32] proposed a method based on the use of an MLP as separator and using as independence criterion the squared error between the output density and a uniform density within a hypercube. The method was shown to be able to extract significant nonlinear features from synthetic-aperture radar images.

In 1998, Yang *et al.* [113] proposed two new nonlinear ICA methods. Both methods used as separator an MLP with a single hidden layer, with a number of hidden units equal to the number of mixture components. One of the methods was a rather direct extension of INFOMAX, being based on the maximization of the output entropy and using output nonlinearities chosen *a priori*, as in INFOMAX.

The other method, instead of minimizing the mutual information indirectly through the output entropy, made a more direct estimation using truncated Gram–Charlier expansions of the extracted components' densities. The Gram–Charlier expansion [39] is an expansion of a pdf "around" a Gaussian with the same variance. Its coefficients depend on the distribution's cumulants that, in this case, were truncated to the fourth order and were estimated from the training set.

Also in 1998, Hochreiter and Schmidhuber [43–45] proposed the so-called LO-COCODE method, which was based on a philosophy similar to (although less complete than) the MDL one used later to develop ensemble learning: it tried to find a low-complexity auto-encoder for the data. Under appropriate conditions, this led the auto-encoder's internal representation of the data to consist of the original sources. The main difference in philosophy, relative to ensemble learning, was that it did not take into account the complexity of the extracted sources and of the modeling error. The method was demonstrated on an artificial nonlinear separation task, both with noiseless and with noisy data.

In 1999, Marques and L. Almeida proposed a method, called *pattern repulsion*, based on a physical analogy with electrostatic repulsion among output patterns [72]. The method was shown to be equivalent to the maximization of the second-order differential Renyi entropy of the output, defined as [112]

$$H_2(\mathbf{Y}) = -\log \int [p(\mathbf{y})]^2 \, d\mathbf{y}.$$

As with Shannon's entropy, this entropy maximization led to a uniform distribution of the outputs within a hypercube, and thus to independent outputs. The method used regularization to handle the indetermination of nonlinear ICA. This method was extended by Hochreiter and Mozer [42], by including both "repulsion" and "attraction," allowing it to deal with nonuniform source distributions. A further theoretical analysis of the method was made in [100].

Also in 1999, Palmieri [80] proposed a method based on the maximization of the output entropy, using as separator an MLP with the restriction that it had, in each hidden layer, the same number of units as the number of mixture components. The possible ill-posedness of the problem was not addressed in this work. The same separator structure (but restricted to two hidden layers) was considered in [73] with a different learning algorithm.

In the same year, Hyvärinen and Pajunen [56] showed that a nonlinear mixture of two independent sources is uniquely separable if the mixture is conformal[25] and the sources' distributions have known, bounded supports.

Still in 1999, Lappalainen (presently Valpola) and Giannakopoulos proposed the ensemble learning method, studied in Section 3.2.2.

In 2000, L. Almeida proposed the MISEP method, studied in Section 3.2.1. Also in 2000, Fyfe and Lai [33] proposed a method that combines the use of kernels with the canonical correlation analysis technique from statistics, and demonstrated the separation of two sinusoids from a nonlinear mixture. The method seems to retain at least some of the indetermination of nonlinear ICA because it does not provide information on which components, among those that are extracted, do correspond to actual sources.

---

[25] A conformal mapping is a mapping that locally preserves orthogonality.

In 2001, Harmeling proposed the kTDSEP method, studied in Section 3.2.3. Also in 2001, Tan *et al.* [99] proposed a method using a radial basis function (RBF) network [39] as a separator. The method used as separation criterion the mutual information of the outputs, together with a matching between moments of the outputs and moments of the sources (the latter moments were assumed to be known *a priori*). The mutual information of the outputs was estimated by expressing the outputs' densities through a truncated Gram–Charlier expansion. The method relied on the smoothing properties of the RBF network, together with the prior knowledge of the sources' moments and the truncation of the Gram–Charlier expansion, to handle the ill-posedness of nonlinear ICA.

Also in 2001, Xiong and Huang [111] proposed an extension of INFOMAX to nonlinear mixtures, using a restricted, truncated power series expansion as a model of the unmixing system.

In 2003, Achard *et al.* [3] proposed a class of quadratic measures of dependence. These measures were only applied to the separation of PNL mixtures.

Also in 2003, Hosseini and Deville proposed a method for the separation of a restricted kind of nonlinear mixture of two sources, of the form $s_i = a_i x_1 + b_i x_2 + c_i x_1 x_2$ [51,52]. They showed that this kind of mixture, in which the nonlinearity appears only through the product of the two sources, is uniquely separable under relatively general conditions. In [52] they proposed a maximum likelihood method for the estimation of the separator's parameters.

Still in 2003, Theis *et al.* [101] proposed a nonlinear ICA method based on geometric concepts. The method was restricted to mixtures that were linear within (hyper-)rings centered on the origin.

In 2004, Blaschke and Wiscott [21] proposed a method that has some aspects in common with kTDSEP: it performs linear ICA in a nonlinearly expanded feature space, and it uses the temporal structure of the signals to address the ill-posedness of nonlinear ICA. However, the method differs from kTDSEP in some important aspects. On the one hand, it performs ICA in the feature space itself, and not in a reduced-dimension intermediate space. Furthermore, it operates directly in the feature space, without relying on the kernel trick. This limits the dimension of the feature space that it can use. On the other hand, the way in which the method uses the signal structure is by employing the principle that the true sources are slower-varying in time than nonlinear combinations or nonlinear distortions of them. The method has been demonstrated in the separation of the "euro mixture" presented in Section 3.2.3. That example used a feature space involving all monomials of degree up to 5, a space that has dimension 20.

Also in 2004, Lee *et al.* [65] proposed a method that is restricted to nonlinear mixtures that are isometric, i.e. such that distances measured in the source space are kept in the mixture space (but the metrics that are used can be non-Euclidean). The restriction to isometric mixtures allows the indetermination of nonlinear ICA to be tackled.

In 2005, the so-called denoising source separation method, originally proposed for linear source separation [87], was extended to nonlinear separation by M. Almeida *et al.*, and was demonstrated on the image separation problem that we examined in Section 3.2.1.4 [12]. The method is not based on an independence criterion. Instead, it uses some prior information about the sources and/or the mixture process to perform a partial separation of the sources. With an application of this partial separation within an appropriate iterative structure, the method is able to achieve a rather complete separation.

As we have seen, a large number of nonlinear source separation methods have been proposed in the literature. Some of them are limited to specific kinds of nonlinear mixtures, either explicitly or due to the restricted kind of separator that they use, but other ones are relatively generic. In our view, this variety of methods reflects both the youth and the difficulty of the topic: it has not stabilized into a well-defined set of methods yet. In Sections 3.2.1 to 3.2.3 we presented in some detail the methods that, in our opinion, have the greatest potential for yielding useful application results.

## 3.3   CONCLUSION

If we compare the results obtained by kTDSEP with those obtained with MISEP, there is a difference that is striking. The kTDSEP method can successfully separate mixtures involving nonlinearities that are much stronger than those that can be handled by MISEP. A good example is the "euro mixture." MISEP, applied to this mixture, yields independent components, but these do not correspond to the original sources. MISEP is not able to deal with the ill-posedness of nonlinear separation in this case, because it uses the assumption that the mixture is mildly nonlinear to be able to perform regularization, and that assumption is not valid in this case.

The question that immediately comes to mind is how can kTDSEP deal with the ill-posedness, even without regularization. The intuition is that its use of the temporal structure of the sources in the separation process greatly reduces the indetermination of nonlinear ICA. There is some evidence to support this idea. In a series of unpublished experiments, jointly performed by Harmeling and L. Almeida, a variant of kTDSEP was used to try to separate a mixture of sources with no time structure (i.e. each of them had independent, identically distributed samples). Of course, TDSEP could not be used for the linear separation step. Therefore another method (INFOMAX) was used for this step. Even though the mixtures that were considered were only mildly nonlinear, all tests failed to recover the original sources. This seems to confirm the idea that the use of temporal structure is what allows kTDSEP to successfully deal with the indetermination. Hosseini and Jutten have analyzed in [53] a number of cases that also suggest that temporal structure can be used for this purpose. We conjecture that temporal (or spatial) structure of the sources will become a very important element in dealing with the indetermination of nonlinear source separation in the future.

CHAPTER 4

# Final Comments

We have made an overview of nonlinear source separation and have studied in some detail the main methods that are currently used to perform this kind of operation. While these methods yield useful results in a number of cases, they are still somewhat hard to apply, and still need much care from the user, both in tuning them and in assessing the quality of the results. Simply put, these methods are still very far from a "black box" use, in which one would just grab a separation routine, supply the data, and immediately get useful results.

The reader would have noticed that there are still few examples of application of nonlinear separation to real-life data. Some people have suggested that there are few naturally occurring nonlinear mixtures. We think this is not so. It is true that many naturally occurring mixtures are approximately linear. This is often the case with acoustic, biomedical, and telecommunications signals, for example, and it is a fortunate situation, because it allows us to deal with them using the well-developed linear separation methods. However, we think that we often do not identify nonlinear mixtures as such, simply because we still do not have powerful enough methods to deal with them. Imagine, for example, a complex set of real-life data such as the stock values from some financial market. We would like to be able to extract the fundamental "sources" that drive these data. These could perhaps be variables such as the investor's confidence, the interest rate, the market liquidity and volatility, and probably also other variables that we are unable to think of. These "sources" are related to the observed data in a nonlinear way. The same could be said about countless other sets of data. If we were able to efficiently perform nonlinear source separation in complex data sets, we would have a very powerful data mining tool, applicable to a very large number of situations. It is therefore worth investing a significant effort into the development of more powerful nonlinear separation methods.

In our view, the main difficulty of nonlinear source separation resides in the ill-posedness of nonlinear ICA, that we have emphasized a number of times throughout this book. While the use of the temporal or spatial structure of signals may provide a significant help in this respect, as illustrated by the kTDSEP method, knowledge about how this structure should be used, and about the capabilities and shortcomings of its use, is still very limited. To our knowledge, the use of signal structure in nonlinear separation is currently limited to kTDSEP, to the method presented in [21], and to ensemble learning with a dynamical model (as in the second example of Section 3.2.2). The experience with these methods is still rather restricted. More work is needed to explore and understand the use of signal structure in addressing the indeterminations of nonlinear ICA.

We would, however, like, to also point in a different direction, as a more fundamental way of addressing the nonlinear source separation problem, and probably also as a way to cope with its indeterminations. Until now, the statistical independence of the sources has been used as the main criterion for separation, both in linear and in nonlinear mixtures. However, sources are not always independent, and in such cases the separation based on independence may be impaired. An example, in the context of linear separation, was given by Pajunen [78]. Another example, within the context of nonlinear separation, was given by L. Almeida in [10]. In [78], Pajunen showed that, while ICA failed to separate the nonindependent sources, a method based on minimizing the complexity of the extracted sources did yield a good separation. That work was just a proof of concept, however, because the method that was used would involve an inordinate amount of computation in any real-life situation. Another work in the same direction is the LOCOCODE method of Hochreiter and Schmidhuber [43–45], that we mentioned in Section 3.2.4. It is a nonlinear separation method that uses as criterion the low complexity of the separator and of its inverse, instead of using independence (LOCOCODE stands for low complexity coding and decoding). A more complete approach in the complexity-based direction is the ensemble learning method that we presented in Section 3.2.2, which is based on the minimum description length principle. This was, to date, and to our knowledge, the most thorough application of complexity-based criteria to nonlinear source separation.

If we think of sources such as speech, images, or other common signals, it is intuitively clear that such sources are, in general, simpler than their mixtures. Furthermore, complexity-based criteria, such as minimum description length, have shown their great potential in several inductive inference situations. Nonlinear source separation can clearly be viewed as an inductive inference problem, in which we wish to infer a set of sources and a mixture model from observed mixture data.

We believe that a fruitful path in nonlinear source separation lies in the use of complexity-based criteria. Their use is not simple, however. For example, a natural complexity measure for images would be their code length when encoded in JPEG with a given quality level. However, this measure is not differentiable, and therefore cannot be used in many of the most efficient optimization methods, which involve the use of first- and/or second-order derivatives. Furthermore, JPEG encoding is somewhat time-consuming, and therefore is not very appropriate for use in an iterative optimization procedure. It is necessary to develop measures of complexity (possibly approximate) which are differentiable and easier to compute. Those measures will probably depend, to a great extent, on the kind of sources to be separated. For example, complexity measures for images are not appropriate for speech and vice versa. In our view, we need to develop methods that are able to exploit elaborate and powerful measures of complexity and that, at the same time, are efficient in computational terms.

The field of nonlinear source separation will keep steadily advancing, with the development of methods that are progressively more powerful, more efficient, easier to use and applicable to a wider range of situations. The benefits to be gained are very large. Linear separation, which is much more advanced today, has already shown the potential of these methods to reveal the "hidden truth" that lies behind complex signals, making them much easier to understand and to process, and giving us access to information that could not be obtained in any other way. A similar potential awaits us as we learn to perform nonlinear separation more effectively.

# Appendix A: Statistical Concepts

## A.1 PASSING A RANDOM VARIABLE THROUGH ITS CUMULATIVE DISTRIBUTION FUNCTION

Consider a continuous random variable $Y$, whose cumulative distribution function, $F_Y$, is continuous, and whose probability density function (pdf) is $p(y)$. Define a new random variable $Z = F_Y(Y)$. We wish to find the distribution of $Z$.

Clearly, the domain of $Z$ is the codomain of $F_Y$, which is the interval $(0, 1)$. Since

$$\frac{\mathrm{d}z}{\mathrm{d}y} = F_Y'(y) = p(y) \geq 0,$$

$F_Y$ is one-to-one wherever $p(y) \neq 0$, and therefore we have, for $z \in (0, 1)$:

$$
\begin{aligned}
p(z) &= p(y) \left| \frac{\mathrm{d}y}{\mathrm{d}z} \right| \\
&= \frac{p(y)}{\mathrm{d}z/\mathrm{d}y} \\
&= \frac{p(y)}{p(y)} \\
&= 1.
\end{aligned}
$$

We conclude that $Z$ will be uniformly distributed in $(0, 1)$, whatever the distribution of the original random variable $Y$. The function $F_Y$ is the nondecreasing function that transforms $Y$ into a random variable which is uniformly distributed in $(0, 1)$. This fact is used by some ICA methods, namely INFOMAX and MISEP.

## A.2 ENTROPY

The entropy of a discrete random variable $X$, as defined by Shannon [29, 91], is

$$
\begin{aligned}
H(X) &= -\sum_i P(x_i) \log P(x_i) \\
&= -\mathrm{E}[\log P(X)],
\end{aligned}
\tag{A.1}
$$

where $P(x_i)$ is the probability that $X$ takes the value $x_i$, and the sum spans all the possible values of $X$. The entropy measures the amount of information that is contained, on average, in $X$. It can also be interpreted as the minimum number of symbols needed to encode $X$, on average.

For example, if we use the base-2 logarithm in (A.1), $H(X)$ represents the minimum number of bits needed, on average, to represent each value of $X$.

Extending this concept to continuous random variables would lead to an infinite entropy. A concept that is often useful is the so-called *differential entropy*, also usually denoted $H(X)$, and defined as

$$H(X) = -\int_{-\infty}^{\infty} p(x)\, \log p(x)\, dx$$
$$= -\mathrm{E}[\log p(X)], \qquad\qquad (A.2)$$

where $p(X)$ is the probability density function of $X$, and the convention $0 \log 0 = 0$ is adopted. The differential entropy does not have the same interpretation as the entropy in terms of coding length, but is useful in relative terms, for comparing different random variables. Its definition extends, in a straightforward way, to multidimensional variables.

Following are two important facts about differential entropy (see [29, Chapter 11] or [91]):

- Among all distributions with support within a given bounded region, the one with the largest differential entropy is the uniform distribution within that region. This is true both for single-dimensional and for multidimensional distributions.

- Among all single-dimensional distributions with zero mean and with a given variance, the one with the largest differential entropy is the Gaussian distribution with the given variance. Among all multidimensional distributions with zero mean and a given covariance matrix, the one with the largest differential entropy is the Gaussian distribution with the given covariance matrix. Among all multidimensional distributions with zero mean and a given variance, the one with the largest differential entropy is the spherically symmetric Gaussian distribution with the given variance.

## A.2.1   Entropy of a Transformed Variable

Assume that we have a multidimensional, continuous random variable $X$ with entropy $H(X)$, and that we obtain from it a new random variable $Z$, through a (possibly nonlinear) invertible transformation $Z = T(X)$. Since the transformation is invertible,

$$p(z) = p(x) \left| \det \frac{\partial x}{\partial z} \right|$$
$$= \frac{p(x)}{|\det J|},$$

where $J = \partial \boldsymbol{z}/\partial \boldsymbol{x}$ is the Jacobian of the transformation $T$. Therefore,

$$
\begin{aligned}
H(\boldsymbol{Z}) &= -\mathrm{E}[\log p(\boldsymbol{z})] \\
&= -\mathrm{E}[\log p(\boldsymbol{x})] + \mathrm{E}[\log |\det J|] \\
&= H(\boldsymbol{X}) + \mathrm{E}[\log |\det J|].
\end{aligned}
$$

## A.3    KULLBACK–LEIBLER DIVERGENCE

The Kullback–Leibler divergence of a density $q$ relative to another density $p$ is defined as [29]

$$
KLD(p, q) = \int p(z) \, \log \frac{p(z)}{q(z)} \, dz,
$$

where the integral extends over the support of $p(z)$. The definition extends, in a straightforward way, to multidimensional densities.

$KLD(p, q)$ is often interpreted as measuring the deviation of an approximate density $q$ from a true density $p$. It is always nonnegative, and is zero if and only if $p = q$. Although it is sometimes called *Kullback–Leibler distance*, it lacks one of the properties of a distance, since it is nonsymmetric: in general, $KLD(p, q) \neq KLD(q, p)$.

## A.4    MUTUAL INFORMATION

Consider a random vector $\boldsymbol{Y}$ whose components $Y_i$ may be mutually dependent. The mutual information of the components of $\boldsymbol{Y}$, that we shall denote by $I(\boldsymbol{Y})$, is the amount of information that is shared by them. It is defined as [29]

$$
I(\boldsymbol{Y}) = \sum_i H(Y_i) - H(\boldsymbol{Y}), \tag{A.3}
$$

where, for the case of continuous random variables, $H$ represents Shannon's differential entropy, and for the case of discrete variables it represents Shannon's entropy.

$I(\boldsymbol{Y})$ is a nonnegative quantity. It is zero if and only if the components $Y_i$ are mutually independent. This agrees with the intuitive concept that we mentioned above: The information shared by the components $Y_i$ is never negative, and is zero only if these components are mutually independent.

$I(\boldsymbol{Y})$ is equal to the Kullback–Leibler divergence between the density of $\boldsymbol{Y}$ and the product of the marginal densities of the components $Y_i$:

$$
I(\boldsymbol{Y}) = KLD\left[ p(\boldsymbol{y}), \prod_i p(y_i) \right].
$$

This is proved, for example, in [29, 39], for the case of a two-dimensional random vector $\boldsymbol{Y}$, and the proof can easily be extended to more than two dimensions.

Since the mutual information $I(Y)$ is zero only if the components of $Y$ are independent and is positive otherwise, its minimization can be used as a criterion for obtaining independent components. Several of the source separation methods studied in this book are based on the minimization of $I(Y)$ as a way to obtain components that are as independent as possible.

An important property of mutual information is that it is not affected by performing invertible, possibly nonlinear, transformations on the individual random variables. More specifically, if we define new random variables $Z_i = \psi_i(Y_i)$ and all the $\psi_i$ are invertible, then $I(Y) = I(Z)$. In fact,

$$I(Z) = \sum_i H(Z_i) - H(Z)$$

$$= \sum_i \left\{ H(Y_i) + \mathrm{E}\left[\log|\psi_i'(Y_i)|\right]\right\} - H(Y) - \mathrm{E}\left[\log\left|\det\frac{\partial Z}{\partial Y}\right|\right]. \qquad (A.4)$$

But the Jacobian $\partial Z/\partial Y$ is a diagonal matrix, its determinant being given by

$$\det\frac{\partial Z}{\partial Y} = \prod_i \psi_i'(Y_i).$$

Therefore,

$$\mathrm{E}\left[\log\left|\det\frac{\partial Z}{\partial Y}\right|\right] = \mathrm{E}\left[\log\prod_i |\psi_i'(Y_i)|\right]$$

$$= \sum_i \mathrm{E}\left[\log|\psi_i'(Y_i)|\right].$$

Consequently the expectations in (A.4) cancel out, and

$$I(Z) = \sum_i H(Y_i) - H(Y)$$

$$= I(Y).$$

# Appendix B: Online Software and Data

## Linear ICA Software

There are many linear ICA software packages and demos available online. These are some sites with pointers to such software:

- ICA Central—`http://www.tsi.enst.fr/icacentral`
- Paris Smaragdis' page—`http://web.media.mit.edu/~paris/ica.html`
- ICA Research Network—`http://www.elec.qmul.ac.uk/icarn/software.html`

## Nonlinear ICA Software

Below are links to some nonlinear ICA software packages and demos available online.

- PNL mixtures—`http://www.lis.inpg.fr/realise_au_lis/demos/sep_sourc/ICAdemo/`
- MISEP—`http://www.lx.it.pt/~lbalmeida/ica/mitoolbox.html`
- Ensemble learning (NFA)—`http://www.cis.hut.fi/projects/bayes/software/`

## Nonlinear Separation Datasets and Dedicated Code

The dataset used for the nonlinear image separation experiments described in Section 3.2.1 and in reference [10] is available at

`http://www.lx.it.pt/~lbalmeida/ica/seethrough/index.html`

The Matlab code used for performing those experiments is available at

`http://www.lx.it.pt/~lbalmeida/ica/seethrough/code/jmlr05/`

# References

[1] S. Achard and C. Jutten, "Identifiability of post nonlinear mixtures," *IEEE Signal Processing Letters*, vol. 12, no. 5, pp. 423–426, 2005. doi:10.1109/LSP.2005.845593

[2] S. Achard, D. Pham, and C. Jutten, "Blind source separation in post nonlinear mixtures," in *Proc. Int. Workshop Independent Component Analysis and Blind Signal Separation*, San Diego, CA, 2001, pp. 295–300. [Online]. Available: http://www-lmc.imag.fr/lmc-sms/Sophie.Achard/Recherche/ICA2001.pdf

[3] S. Achard, D. Pham, and C. Jutten, "Quadratic dependence measure for nonlinear blind sources separation," in *Proc. Int. Workshop Independent Component Analysis and Blind Signal Separation*, Nara, Japan, 2003. [Online]. Available: http://www.kecl.ntt.co.jp/icl/signal/ica2003/cdrom/data/0098.pdf

[4] L. Almeida, "Multilayer perceptrons," in *Handbook of Neural Computation*, E. Fiesler and R. Beale, Eds. Bristol, U.K.: Institute of Physics. 1997. [Online]. Available: http://www.lx.it.pt/~lbalmeida/papers/AlmeidaHNC.pdf

[5] L. Almeida, "Linear and nonlinear ICA based on mutual information," in *Proc. Symp. 2000 Adaptive Systems for Signal Processing, Communications, and Control*, Lake Louise, Alberta, Canada, 2000. [Online]. Available: http://www.lx.it.pt/~lbalmeida/papers/AlmeidaASSPCC00.ps.zip

[6] L. Almeida, "Simultaneous MI-based estimation of independent components and of their distributions," in *Proc. Second Int. Workshop Independent Component Analysis and Blind Signal Separation*, Helsinki, Finland, 2000, pp. 169–174. [Online]. Available: http://www.lx.it.pt/~lbalmeida/papers/AlmeidaICA00.ps.zip

[7] L. Almeida, "Faster training in nonlinear ICA using MISEP," in *Proc. Int. Workshop Independent Component Analysis and Blind Signal Separation*, Nara, Japan, 2003, pp. 113–118. [Online]. Available: http://www.lx.it.pt/~lbalmeida/papers/AlmeidaICA03.pdf

[8] L. Almeida, "MISEP—Linear and nonlinear ICA based on mutual information," *Journal of Machine Learning Research*, vol. 4, pp. 1297–1318, 2003. [Online]. Available: http://www.jmlr.org/papers/volume4/almeida03a/almeida03a.pdf

[9] L. Almeida, "Linear and nonlinear ICA based on mutual information—the MISEP method," *Signal Processing*, vol. 84, no. 2, pp. 231–245, 2004. [Online]. Available: http://www.lx.it.pt/~lbalmeida/papers/AlmeidaSigProc03.pdf doi:10.1016/j.sigpro.2003.10.008

[10]   L. Almeida, "Separating a real-life nonlinear image mixture," *Journal of Machine Learning Research*, vol. 6, pp. 1199–1229, July 2005. [Online]. Available: http://www.jmlr.org/papers/volume4/almeida03a/almeida03a.pdf

[11]   L. Almeida and M. Faria, "Separating a real-life nonlinear mixture of images," in *Proc. Int. Workshop Independent Component Analysis and Blind Signal Separation*, Series Lecture Notes in Artificial Intelligence, no. 3195, C. G. Puntonet and A. Prieto, Eds. New York, NY: Springer-Verlag, 2004, pp. 729–736. [Online]. Available: http://www.lx.it.pt/~lbalmeida/papers/AlmeidaICA04.pdf

[12]   M. Almeida, H. Valpola, and J. Särelä, "Separation of nonlinear image mixtures by denoising source separation," in *Proc. Int. Conf. Independent Component Analysis and Blind Signal Separation*, accepted for publication. [Online]. Available: http://www.lce.hut.fi/~harri/publications/ICA06final.pdf

[13]   S.-I. Amari, "Natural gradient works efficiently in learning," *Neural Computation*, vol. 10, no. 2, pp. 252–276, 1998.doi:10.1162/089976698300017746

[14]   M. Babaie-Zadeh, C. Jutten, and K. Nayebi, "Separating convolutive post-nonlinear mixtures," in *Proc. Int. Conf. Independent Component Analysis and Blind Source Separation*, San Diego, CA, 2001, pp. 138–143.

[15]   M. Babaie-Zadeh, C. Jutten, and K. Nayebi, "A geometric approach for separating post nonlinear mixtures," in *Proc. XI Eur. Sig. Proc. Conf.*, Toulouse, France, 2002, pp. 11–14.

[16]   M. Babaie-Zadeh, C. Jutten, and K. Nayebi, "Minimization-projection (MP) approach for blind source separation in different mixing models," in *Proc. Fourth Int. Symp. Independent Component Analysis and Blind Signal Separation*, Nara, Japan, 2003, pp. 915–920.

[17]   M. Babaie-Zadeh, C. Jutten, and K. Nayebi, "A minimization-projection (MP) approach for blind separating convolutive mixtures," in *Proc. IEEE Int. Conf. Acoustics, Speech and Signal Processing*, Montreal, Canada, 2004, pp. 533–536.

[18]   F. Bach and M. Jordan, "Kernel independent component analysis," *Journal of Machine Learning Research*, vol. 3, pp. 1–48, 2002.doi:10.1162/153244303768966085

[19]   A. Bell and T. Sejnowski, "An information-maximization approach to blind separation and blind deconvolution," *Neural Computation*, vol. 7, pp. 1129–1159, 1995. [Online]. Available: ftp://ftp.cnl.salk.edu/pub/tony/bell.blind.ps

[20]   A. Belouchrani, K. Meraim, J.-F. Cardoso, and E. Moulines, "A blind source separation technique based on second order statistics," *IEEE Transactions on Signal Processing*, vol. 45, no. 2, pp. 434–444, 1997. [Online]. Available: http://www.tsi.enst.fr/~cardoso/Papers.PDF/ieeesobi.pdfdoi:10.1109/78.554307

[21]   T. Blaschke and L. Wiskott, "Independent slow feature analysis and nonlinear blind source separation," in *Proc. Int. Workshop Independent Component Analysis and Blind Signal*

*Separation*, Series Lecture Notes in Artificial Intelligence, no. 3195, C. G. Puntonet and A. Prieto, Eds. Springer-Verlag, 2004, pp. 742–749. [Online]. Available: http://itb .biologie.hu-berlin.de/~blaschke/publications/isfa.pdf

[22] R. Boscolo, H. Pan, and V. Roychowdhury, "Independent component analysis based on nonparametric density estimation," *IEEE Transactions on Neural Networks*, vol. 15, no. 1, pp. 55–65, January 2004. [Online]. Available: http://www.ee.ucla.edu/faculty/ papers/vwani_trans-neural_jan04.pdf doi:10.1109/TNN.2003.820667

[23] G. Burel, "Blind separation of sources: A nonlinear neural algorithm," *Neural Networks*, vol. 5, no. 6, pp. 937–947, 1992.doi:10.1016/S0893-6080(05)80090-5

[24] C. Burges, "A tutorial on support vector machines for pattern recognition," *Data Mining and Knowledge Discovery*, vol. 2, no. 2, pp. 121–167, 1998. [Online]. Available: http://www.kernel-machines.org/papers/Burges98.ps.gz doi:10.1023/A:1009715923555

[25] J.-F. Cardoso, "The invariant approach to source separation," in *Proc. NOLTA*, 1995, pp. 55–60. [Online]. Available: http://www.tsi.enst.fr/~cardoso/Papers.PDF/nolta95 .pdf

[26] J.-F. Cardoso and A. Souloumiac, "Blind beamforming for non Gaussian signals," *IEE Proceedings-F*, vol. 140, no. 6, pp. 362–370, 1993. [Online]. Available: http://www.tsi .enst.fr/~cardoso/Papers.PDF/iee.pdf

[27] A. Cichocki and S.-I. Amari, *Adaptive Blind Signal and Image Processing—Learning Algorithms and Applications*. New York, NY: Wiley, 2002.

[28] P. Comon, "Independent component analysis—a new concept?" *Signal Processing*, vol. 36, pp. 287–314, 1994.doi:10.1016/0165-1684(94)90029-9

[29] T. M. Cover and J. A. Thomas, *Elements of Information Theory*. New York, NY: Wiley, 1991.

[30] G. Darmois, "Analyse générale des liaisons stochastiques," *Rev. Inst. Internat. Stat.*, vol. 21, pp. 2–8, 1953.

[31] G. Deco and W. Brauer, "Nonlinear higher-order statistical decorrelation by volume-conserving neural architectures," *Neural Networks*, vol. 8, pp. 525–535, 1995. doi:10.1016/0893-6080(94)00108-X

[32] J. Fisher and J. Principe, "Entropy manipulation of arbitrary nonlinear mappings," in *Proc. IEEE Workshop Neural Networks for Signal Processing*, Amelia Island, FL, 1997, pp. 14–23. [Online]. Available: http://www.cnel.ufl.edu/bib/pdf_papers/fisher_nnsp97.pdf

[33] C. Fyfe and P. Lai, "ICA using kernel canonical correlation analysis," in *Proc. Int. Workshop Independent Component Analysis and Blind Signal Separation*, Helsinki, Finland, 2000, pp. 279–284. [Online]. Available: http://www.cis.hut.fi/ica2000/proceedings/ 0279.pdf

[34] P. Grünwald, "A tutorial introduction to the minimum description length principle," in *Advances in Minimum Description Length: Theory and Applications*, P. Grünwald, I. Myung, and M. Pitt, Eds. Cambridge, MA: MIT Press, 2005. [Online]. Available: http://www.cwi.nl/~pdg/ftp/mdlintro.pdf

[35] M. Hansen and B. Yu, "Model selection and the principle of minimum description length," *Journal of American Statistical Association*, vol. 96, pp. 746–774, 2001. [Online]. Available: http://www.stat.ucla.edu/~cocteau/papers/pdf/mdl.pdf doi:10.1198/016214501753168398

[36] M. Haritopoulos, H. Yin, and N. Allinson, "Image denoising using SOM-based nonlinear independent component analysis," *Neural Networks*, vol. 15, no. 8–9, pp. 1085–1098, 2002. doi:10.1016/S0893-6080(02)00081-3

[37] S. Harmeling, A. Ziehe, M. Kawanabe, B. Blankertz, and K.-R. Müller, "Nonlinear blind source separation using kernel feature spaces," in *Proc. Int. Workshop Independent Component Analysis and Blind Signal Separation*, T.-W. Lee, Ed., San Diego, CA, 2001, pp. 102–107. [Online]. Available: http://ica2001.ucsd.edu/index_files/pdfs/080-harmeling.pdf

[38] S. Harmeling, A. Ziehe, M. Kawanabe, and K.-R. Müller, "Kernel-based nonlinear blind source separation," *Neural Computation*, vol. 15, pp. 1089–1124, 2003. [Online]. Available: http://ida.first.fraunhofer.de/~harmeli/papers/article_on_ktdsep.pdf doi:10.1162/089976603765202677

[39] S. Haykin, *Neural Networks—A Comprehensive Foundation*, 2nd ed. Upper Saddle River, NJ: Prentice-Hall, 1999.

[40] S. Haykin and J. Principe, "Making sense of a complex world," *IEEE Signal Processing Magazine*, vol. 15, no. 3, pp. 66–81, May 1998. [Online]. Available: http://www.cnel.ufl.edu/alltest.php?type=journals&id=5 doi:10.1109/79.671132

[41] R. Hecht-Nielsen, "Replicator neural networks for universal optimal source coding," *Science*, vol. 269, no. 5232, pp. 1860–1863, 1995.

[42] S. Hochreiter and M. C. Mozer, "An electric field approach to independent component analysis," in *Proc. Second Int. Workshop Independent Component Analysis and Blind Signal Separation*, Helsinki, Finland, 2000, pp. 45–50. [Online]. Available: http://www.cis.hut.fi/ica2000/proceedings/0045.pdf

[43] S. Hochreiter and J. Schmidhuber, "LOCOCODE versus PCA and ICA," in *Proc. Int. Conf. Artificial Neural Networks*, Sweden, 1998, pp. 669–674. [Online]. Available: ftp://ftp.idsia.ch/pub/juergen/icann98.ps.gz

[44] S. Hochreiter and J. Schmidhuber, "Feature extraction through LOCOCODE," *Neural Computation*, vol. 11, no. 3, pp. 679–714, 1999. [Online]. Available: ftp://ftp.idsia.ch/pub/juergen/lococode.pdf doi:10.1162/089976699300016629

[45]  S. Hochreiter and J. Schmidhuber, "LOCOCODE performs nonlinear ICA without knowing the number of sources," in *Proc. First Int. Workshop Independent Component Analysis and Signal Separation*, J. F. Cardoso, C. Jutten, and P. Loubaton, Eds., Aussois, France, 1999, pp. 277–282.

[46]  A. Honkela, "Speeding up cyclic update schemes by pattern searches," in *Proc. Ninth Int. Conf. Neural Information Processing*, Singapore, 2002, pp. 512–516.

[47]  A. Honkela, S. Harmeling, L. Lundqvist, and H. Valpola, "Using kernel PCA for initialisation of variational Bayesian nonlinear blind source separation method," in *Proc. Int. Workshop Independent Component Analysis and Blind Signal Separation*, Granada, Spain, 2004, pp. 790–797.

[48]  A. Honkela and H. Valpola, "Variational learning and bits-back coding: An information-theoretic view to Bayesian learning," *IEEE Transactions on Neural Networks*, vol. 15, no. 4, pp. 800–810, 2004.doi:10.1109/TNN.2004.828762

[49]  A. Honkela and H. Valpola, "Unsupervised variational bayesian learning of nonlinear models," in *Advances in Neural Information Processing Systems*, vol. 17, L. K. Saul, Y. Weis, and L. Bottou, Eds., 2005, pp. 593–600. [Online]. Available: http://books.nips.cc/papers/files/nips17/NIPS2004_0322.pdf

[50]  A. Honkela, H. Valpola, and J. Karhunen, "Accelerating cyclic update algorithms for parameter estimation by pattern searches," *Neural Processing Letters*, vol. 17, no. 2, pp. 191–203, 2003.doi:10.1023/A:1023655202546

[51]  S. Hosseini and Y. Deville, "Blind separation of linear-quadratic mixtures of real sources," in *Proc. IWANN*, vol. 2, Mao, Menorca, Spain, 2003, pp. 241–248.

[52]  S. Hosseini and Y. Deville, "Blind maximum likelihood separation of a linear-quadratic mixture," in *Proc. Int. Workshop Independent Component Analysis and Blind Signal Separation*, Series Lecture Notes in Artificial Intelligence, no. 3195. Springer-Verlag, 2004. [Online]. Available: http://webast.ast.obs-mip.fr/people/ydeville/papers/ica04_1.pdf

[53]  S. Hosseini and C. Jutten, "On the separability of nonlinear mixtures of temporally correlated sources," *IEEE Signal Processing Letters*, vol. 10, no. 2, pp. 43–46, February 2003.doi:10.1109/LSP.2002.807871

[54]  A. Hyvärinen, "Fast and robust fixed-point algorithms for independent component analysis," *IEEE Transactions on Neural Networks*, vol. 10, no. 3, pp. 626–634, 1999. [Online]. Available: http://www.cs.helsinki.fi/u/ahyvarin/papers/TNN99new.pdf doi:10.1109/72.761722

[55]  A. Hyvärinen, J. Karhunen, and E. Oja, *Independent Component Analysis*. New York, NY: Wiley, 2001.

[56]  A. Hyvärinen and P. Pajunen, "Nonlinear independent component analysis: Existence and uniqueness results," *Neural Networks*, vol. 12, no. 3, pp. 429–439, 1999.

[Online]. Available: http://www.cis.hut.fi/~aapo/ps/NN99.psdoi:10.1016/S0893-6080(98)00140-3

[57]   A. Ilin and A. Honkela, "Post-nonlinear independent component analysis by variational Bayesian learning," in *Proc. Int. Conf. Independent Component Analysis and Blind Source Separation*, Granada, Spain, 2004, pp. 766–773.

[58]   A. Iline, H. Valpola, and E. Oja, "Detecting process state changes by nonlinear blind source separation," in *Proc. Int. Workshop Independent Component Analysis and Blind Signal Separation*, San Diego, CA, 2001, pp. 710–715. [Online]. Available: http://www.cis.hut.fi/harri/papers/ICAalex.ps.gz

[59]   C. Jutten and J. Karhunen, "Advances in blind source separation (BSS) and independent component analysis (ICA) for nonlinear mixtures," *International Journal of Neural Systems*, vol. 14, no. 5, pp. 267–292, 2004. [Online]. Available: http://www.worldscinet.com/128/14/preserved-docs/1405/S012906570400208X.pdf doi:10.1142/S012906570400208X

[60]   J. Karvanen and T. Tanaka, "Temporal decorrelation as preprocessing for linear and post-nonlinear ICA," in *Proc. Int. Conf. Independent Component Analysis and Blind Source Separation*, Granada, Spain, 2004, pp. 774–781.

[61]   H. Lappalainen and X. Giannakopoulos, "Multi-layer perceptrons as nonlinear generative models for unsupervised learning: A Bayesian treatment," in *Proc. ICANN*, Edinburgh, Scotland, 1999, pp. 19–24. [Online]. Available: http://www.cis.hut.fi/harri/icann99.ps.gz

[62]   H. Lappalainen and A. Honkela, "Bayesian nonlinear independent component analysis by multi-layer perceptrons," in *Advances in Independent Component Analysis*, M. Girolami, Ed. Springer-Verlag, 2000, pp. 93–121. [Online]. Available: http://www.cis.hut.fi/harri/ch7.ps.gz

[63]   H. Lappalainen, A. Honkela, X. Giannakopoulos, and J. Karhunen, "Nonlinear source separation using ensemble learning and MLP networks," in *Proc. Symp. 2000 Adaptive Systems for Signal Processing, Communications, and Control*, Lake Louise, Alberta, Canada, October 1–4 2000, pp. 187–192. [Online]. Available: http://www.cis.hut.fi/projects/ica/bayes/papers/llouise.ps.gzdoi:full_text

[64]   Y. LeCun, I. Kanter, and S. Solla, "Eigenvalues of covariance matrices: Application to neural-network learning," *Physical Review Letters*, vol. 66, no. 18, pp. 2396–2399, 1991. [Online]. Available: http://yann.lecun.com/exdb/publis/pdf/lecun-kanter-solla-91.pdfdoi:10.1103/PhysRevLett.66.2396

[65]   J. Lee, C. Jutten, and M. Verleysen, "Non-linear ICA by using isometric dimensionality reduction," in *Proc. Int. Workshop Independent Component Analysis and Blind Signal Separation*, Series Lecture Notes in Artificial Intelligence, no. 3195.

Springer-Verlag, 2004, pp. 710–717. [Online]. Available: http://www.dice.ucl.ac.be/~verleyse/papers/ica04jl.pdf

[66] T.-W. Lee, M. Girolami, and T. Sejnowski, "Independent component analysis using an extended infomax algorithm for mixed sub-gaussian and super-gaussian sources," *Neural Computation*, vol. 11, pp. 417–441, 1999.doi:10.1162/089976699300016719

[67] T. W. Lee, B. Koehler, and R. Orglmeister, "Blind source separation of nonlinear mixing models," in *Proc. Neural Networks for Signal Processing*, 1997, pp. 406–415. [Online]. Available: http://www.cnl.salk.edu/~tewon/Public/nnsp97.ps.gz

[68] J. Lin, D. Grier, and J. Cowan, "Source separation and density estimation by faithful equivariant SOM," in *Advances in Neural Information Processing Systems*. Cambridge, MA: MIT Press, 1997, pp. 536–542.

[69] J. Lin, D. Grier, and J. Cowan, "Faithful representation of separable input distributions," *Neural Computation*, vol. 9, pp. 1305–1320, 1997.

[70] E. Lorenz, "Deterministic nonperiodic flow," *Journal of Atmospheric Sciences*, vol. 20, pp. 130–141, 1963.doi:10.1175/1520-0469(1963)020<0130:DNF>2.0.CO;2

[71] G. Marques and L. Almeida, "An objective function for independence," in *Proc. Int. Conf. Neural Networks*, Washington, DC, 1996, pp. 453–457.

[72] G. Marques and L. Almeida, "Separation of nonlinear mixtures using pattern repulsion," in *Proc. First Int. Workshop Independent Component Analysis and Signal Separation*, J. F. Cardoso, C. Jutten, and P. Loubaton, Eds., Aussois, France, 1999, pp. 277–282. [Online]. Available: http://www.lx.it.pt/~lbalmeida/papers/MarquesAlmeidaICA99.ps.zip

[73] R. Martín-Clemente, S. Hornillo-Mellado, J. Acha, F. Rojas, and C. Puntonet, "MLP-based source separation for MLP-like nonlinear mixtures," in *Proc. Int. Workshop Independent Component Analysis and Blind Signal Separation*, Nara, Japan, 2003. [Online]. Available: http://www.kecl.ntt.co.jp/icl/signal/ica2003/cdrom/data/0114.pdf

[74] T. Mitchell, *Machine Learning*. New York, NY: McGraw Hill, 1997.

[75] L. Molgedey and H. Schuster, "Separation of a mixture of independent signals using time delayed correlations," *Physical Review Letters*, vol. 72, pp. 3634–3636, 1994. [Online]. Available: http://www.theo-physik.uni-kiel.de/thesis/molgedey94.ps.gz doi:10.1103/PhysRevLett.72.3634

[76] K.-R. Müller, S. Mika, G. Rätsch, K. Tsuda, and B. Schölkopf, "An introduction to kernel-based learning algorithms," *IEEE Transactions on Neural Networks*, vol. 12, no. 2, pp. 181–201, May 2001. [Online]. Available: http://mlg.anu.edu.au/~raetsch/ps/review.pdfdoi:10.1109/72.914517

[77] P. Pajunen, "Nonlinear independent component analysis by self-organizing maps," in *Artificial Neural Networks—ICANN 96, Proc. 1996 International Conference on Artificial Neural Networks*, Bochum, Germany, 1996, pp. 815–819.

[78]  P. Pajunen, "Blind source separation of natural signals based on approximate complexity minimization," in *Proc. Int. Workshop Independent Component Analysis and Blind Signal Separation*, Aussois, France, January 1999, pp. 267–270. [Online]. Available: http://www.cis.hut.fi/~ppajunen/papers/ica99_second.ps

[79]  P. Pajunen and J. Karhunen, "A maximum likelihood approach to nonlinear blind source separation," in *Proc. Int. Conf. Artificial Neural Networks*, Lausanne, Switzerland, October 1997, pp. 541–546. [Online]. Available: http://www.cis.hut.fi/~ppajunen/papers/icann97.ps.gz

[80]  F. Palmieri, D. Mattera, and A. Budillon, "Multi-layer independent component analysis (MLICA)," in *Proc. First Int. Workshop Independent Component Analysis and Signal Separation*, J. F. Cardoso, C. Jutten, and P. Loubaton, Eds., Aussois, France, 1999, pp. 93–97.

[81]  L. Parra, "Symplectic nonlinear independent component analysis," in *Advances in Neural Information Processing Systems 8*, D. Touretzky, M. Mozer, and M. Hasselmo, Eds., Cambridge, MA: MIT Press 1996, pp. 437–443. [Online]. Available: http://newton.bme.columbia.edu/~lparra/publish/Parra.NIPS95.pdf

[82]  L. Parra, G. Deco, and S. Miesbach, "Statistical independence and novelty detection with information preserving nonlinear maps," *Neural Computation*, vol. 8, pp. 260–269, 1996. [Online]. Available: http://newton.bme.columbia.edu/~lparra/publish/nc96.pdf

[83]  B. Pearlmutter and L. Parra, "Maximum likelihood blind source separation: A context-sensitive generalization of ICA," in *Advances in Neural Information Processing Systems*. Cambridge, MA: MIT press, 1997, pp. 613–619. [Online]. Available: http://newton.bme.columbia.edu/~lparra/publish/nips96.pdf

[84]  B. Ripley, *Pattern Recognition and Neural Networks*. Cambridge, UK: Cambridge University Press, 1996.

[85]  J. Rissanen, "Modeling by shortest data description," *Automatica*, vol. 14, pp. 465–471, 1978.doi:10.1016/0005-1098(78)90005-5

[86]  F. Rojas, I. Rojas, R. Clemente, and C. Puntonet, "Nonlinear blind source separation using genetic algorithms," in *Proc. Int. Workshop Independent Component Analysis and Blind Signal Separation*, San Diego, CA, 2001. [Online]. Available: http://ica2001.ucsd.edu/index_files/pdfs/030-rojas.pdf

[87]  J. Särelä and H. Valpola, "Denoising source separation," *Journal of Machine Learning Research*, vol. 6, pp. 233–272, 2005.

[88]  J. Schmidhuber, "Learning factorial codes by predictability minimization," *Neural Computation*, vol. 4, no. 6, pp. 863–879, 1992.

[89]  B. Schölkopf and A. Smola, *Learning with Kernels*. Cambridge, MA: MIT Press, 2002.

[90] B. Schölkopf, A. Smola, and K.-R. Müller, "Nonlinear component analysis as a kernel eigenvalue problem," *Neural Computation*, vol. 10, pp. 1299–1319, 1998. [Online]. Available: http://users.rsise.anu.edu.au/~smola/papers/SchSmoMul98.pdf doi:10.1162/089976698300017467

[91] C. Shannon, "A mathematical theory of communication," *Bell System Technical Journal*, vol. 27, pp. 379–423 and 623–656, July and October 1948. [Online]. Available: http://cm.bell-labs.com/cm/ms/what/shannonday/shannon1948.pdf

[92] J. Solé, C. Jutten, and D. T. Pham, "Fast approximation of nonlinearities for improving inversion algorithms of pnl mixtures and Wiener systems," *Signal Processing*, 2004.

[93] J. Solé, C. Jutten, and A. Taleb, "Parametric approach to blind deconvolution of nonlinear channels," *Neurocomputing*, vol. 48, pp. 339–355, 2002.doi:10.1016/S0925-2312(01)00651-8

[94] A. Taleb and C. Jutten, "Entropy optimization, application to blind source separation," in *Proc. Int. Conf. Artificial Neural Networks*, Lausanne, Switzerland, 1997, pp. 529–534.

[95] A. Taleb and C. Jutten, "Nonlinear source separation: The post-nonlinear mixtures," in *Proc. 1997 Eur. Symp. Artifcial Neural Networks*, Bruges, Belgium, 1997, pp. 279–284.

[96] A. Taleb and C. Jutten, "Batch algorithm for source separation in post-nonlinear mixtures," in *Proc. First Int. Workshop Independent Component Analysis and Signal Separation*, Aussois, France, 1999, pp. 155–160.

[97] A. Taleb and C. Jutten, "Source separation in post-nonlinear mixtures," *IEEE Transactions on Signal Processing*, vol. 47, pp. 2807–2820, 1999.doi:10.1109/78.790661

[98] A. Taleb, J. Solé, and C. Jutten, "Quasi-nonparametic blind inversion of Wiener systems," *IEEE Transactions on Signal Processing*, vol. 49, no. 5, pp. 917–924, 2001. doi:10.1109/78.917796

[99] Y. Tan, J. Wang, and J. Zurada, "Nonlinear blind source separation using a radial basis function network," *IEEE Transactions on Neural Networks*, vol. 12, no. 1, pp. 124–134, 2001.doi:10.1109/72.896801

[100] F. Theis, C. Bauer, C. Puntonet, and E. Lang, "Pattern repulsion revisited," in *Proc. IWANN*, Series Lecture Notes in Computer Science, no. 2085. New York, NY: Springer-Verlag, 2001, pp. 778–785. [Online]. Available: http://homepages.uni-regensburg.de/~thf11669/publications/theis01patternrep_IWANN01.pdf

[101] F. Theis, C. Puntonet, and E. Lang, "Nonlinear geometric ICA," in *Proc. Int. Workshop Independent Component Analysis and Blind Signal Separation*, Nara, Japan, 2003, pp. 275–280. [Online]. Available: http://homepages.uni-regensburg.de/~thf11669/publications/theis03nonlineargeo_ICA03.pdf

[102]    H. Valpola, "Bayesian ensemble learning for nonlinear factor analysis," *Acta Polytechnica Scandinavica, Mathematics and Computing Series*, no. 108, 2000. [Online]. Available: http://www.cis.hut.fi/harri/thesis/

[103]    H. Valpola, "Nonlinear independent component analysis using ensemble learning: Experiments and discussion," in *Proc. Second Int. Workshop Independent Component Analysis and Blind Signal Separation*, Helsinki, Finland, 2000, pp. 351–356.

[104]    H. Valpola, "Nonlinear independent component analysis using ensemble learning: Theory," in *Proc. Second Int. Workshop Independent Component Analysis and Blind Signal Separation*, Helsinki, Finland, 2000, pp. 251–256.

[105]    H. Valpola, M. Harva, and J. Karhunen, "Hierarchical models of variance sources," *Signal Processing*, vol. 84, no. 2, pp. 267–282, 2004.doi:10.1016/j.sigpro.2003.10.014

[106]    H. Valpola and J. Karhunen, "An unsupervised ensemble learning method for nonlinear dynamic state-space models," *Neural Computation*, vol. 14, no. 11, pp. 2647–2692, 2002. [Online]. Available: http://www.cis.hut.fi/harri/papers/ValpolaNC02.pdf doi:10.1162/089976602760408017

[107]    H. Valpola, T. Östman, and J. Karhunen, "Nonlinear independent factor analysis by hierarchical models," in *Proc. Int. Workshop Independent Component Analysis and Blind Signal Separation*, Nara, Japan, 2003, pp. 257–262. [Online]. Available: http://www.cis.hut.fi/harri/papers/ica2003_hnfa.pdf

[108]    H. Valpola, T. Raiko, and J. Karhunen, "Building blocks for hierarchical latent variable models," in *Proc. Int. Workshop Independent Component Analysis and Blind Signal Separation*, San Diego, CA, 2001, pp. 716–721. [Online]. Available: http://www.cis.hut.fi/harri/papers/ica2001.ps.gz

[109]    N. Vlassis, "Efficient source adaptivity in independent component analysis," *IEEE Transactions on Neural Networks*, vol. 12, no. 3, pp. 559–566, May 2001. doi:10.1109/72.925558

[110]    C. Wallace, "Classification by minimum-message-length inference," in *Proc. Advances in Computing and Information—ICCI 90*, vol. 468, Berlin, Germany, S. G. Aki, F. Fiala, and W. W. Koczkodaj, Eds., 1990, pp. 72–81.

[111]    Z. Xiong and T. Huang, "Nonlinear independent component analysis (ICA) using power series and application to blind source separation," in *Proc. Int. Workshop Independent Component Analysis and Blind Signal Separation*, San Diego, CA, 2001. [Online]. Available: http://ica2001.ucsd.edu/index_files/pdfs/031-xiong.pdf

[112]    D. Xu, J. Principe, J. Fisher, and H.-C. Wu, "A novel measure for independent component analysis," in *Proc. ICASSP*, vol. 2, Seattle, 1998.

[113]    H. Yang, S.-I. Amari, and A. Clichocki, "Information theoretic approach to blind separation of sources in non-linear mixture," *Signal Processing*, vol. 64, no. 3, pp. 291–300, 1998.doi:10.1016/S0165-1684(97)00196-5

[114]   A. Ziehe, M. Kawanabe, S. Harmeling, and K.-R. Müller, "Separation of post-nonlinear mixtures using ACE and temporal decorrelation," in *Proc. Int. Conf. Independent Component Analysis and Blind Source Separation*, San Diego, CA, 2001, pp. 433–438.

[115]   A. Ziehe, M. Kawanabe, S. Harmeling, and K.-R. Müller, "Blind separation of post-nonlinear mixtures using gaussianizing transformations and temporal decorrelation," in *Proc. Int. Workshop Independent Component Analysis and Blind Signal Separation*, Nara, Japan, 2003, pp. 269–274. [Online]. Available: http://www.kecl.ntt.co.jp/icl/signal/ica2003/cdrom/data/0208.pdf

[116]   A. Ziehe, M. Kawanabe, S. Harmeling, and K.-R. Müller, "Blind separation of post-nonlinear mixtures using linearizing transformations and temporal decorrelation," *Journal of Machine Learning Research*, vol. 4, pp. 1319–1338, December 2003. [Online]. Available: http://www.jmlr.org/papers/volume4/ziehe03a/ziehe03a.pdf

[117]   A. Ziehe and K.-R. Müller, "TDSEP—An efficient algorithm for blind separation using time structure," in *Proc. Int. Conf. Artificial Neural Networks*, Skövde, Sweden, 1998, pp. 675–680. [Online]. Available: http://wwwold.first.gmd.de/persons/Mueller.Klaus-Robert/ICANN_tdsep.ps.gz

# Biography

**Luis B. Almeida** was born in 1950 in Lisbon, Portugal. He graduated in Electrical Engineering from the Instituto Superior Técnico, Lisbon, in 1972, and obtained the PhD and "Agregado" degrees from the Technical University of Lisbon in 1983 and in 1988 respectively. He has been teaching at the Instituto Superior Técnico since 1970, and is a full professor since 1995. Currently he teaches courses on Signals and Systems and on Neural Networks and Machine Learning.

Luis B. Almeida is a full professor of Signals and Systems, and of Neural Networks and Machine Learning, at Instituto Superior Técnico, Technical University of Lisbon, and a researcher at the Telecommunications Institute, Lisbon, Portugal. He holds a Ph. D. in Signal Processing from the Technical University of Lisbon. He has formerly taught Systems Theory, Telecommunications, Digital Systems and Mathematical Analysis, among others.

Luis B. Almeida's current research focuses on nonlinear source separation. Formerly he has performed research on speech modeling and coding, Fourier and time-frequency analysis of signals, and training algorithms for neural networks. Some highlights of his work include the sinusoidal model for voiced speech, currently in use in INMARSAT and IRIDIUM telephones (developed with F.M. Silva and J.S. Marques), work on the Fractional Fourier Transform, the development of recurrent backpropagation, and the development of the MISEP method of nonlinear source separation.

Luis B. Almeida has been a founding Vice-President of the European Neural Network Society and the founding President of INESC-ID (a nonprofit research institute associated with the Technical University of Lisbon).